THE MATERNAL IMPRINT

The Maternal Imprint

THE CONTESTED SCIENCE OF
MATERNAL-FETAL EFFECTS

Sarah S. Richardson

The University of Chicago Press   CHICAGO AND LONDON

The University of Chicago Press, Chicago 60637
The University of Chicago Press, Ltd., London
© 2021 by Sarah S. Richardson
Published 2021.
Printed in the United States of America

30 29 28 27 26 25 24 23 22 21   1 2 3 4 5

ISBN-13: 978-0-226-54477-9 (cloth)
ISBN-13: 978-0-226-54480-9 (paper)
ISBN-13: 978-0-226-80707-2 (e-book)
DOI: https://doi.org/10.7208/chicago/9780226807072.001.0001

Published with the support of the Susan E. Abrams Fund

Library of Congress Cataloging-in-Publication Data
Names: Richardson, Sarah S., 1980– author.
Title: The maternal imprint : the contested science of maternal-fetal effects /
   Sarah S. Richardson.
Description: Chicago : University of Chicago Press, 2021. |
   Includes bibliographical references and index.
Identifiers: LCCN 2021012763 | ISBN 9780226544779 (cloth) |
   ISBN 9780226544809 (paperback) | ISBN 9780226807072 (e-book)
Subjects: LCSH: Fetus—Development. | Maternal-fetal exchange. |
   Mother and infant.
Classification: LCC RG613.R53 2021 | DDC 612.6/47—dc23
LC record available at https://lccn.loc.gov/2021012763

♾ This paper meets the requirements of ANSI/NISO Z39.48-1992
(Permanence of Paper).

For Ace

Contents

CHAPTER 1

Introduction:
The Maternal Imprint

This book is about the bewitching idea that the environment in which
you are gestated leaves a permanent imprint on you and your future
descendants. I first became intrigued by this idea a decade ago, when
I encountered neuroscientist Rachel Yehuda's studies of intergenerational Holocaust trauma in the families of Jewish survivors. Children of
Holocaust survivors experience higher rates of vulnerability to trauma
themselves, and Yehuda believes that this is because they were gestated
in an environment with high cortisol, a hormone critical in stress regulation. This intrauterine experience, she contends, permanently modified survivors' children's genome regulation so that as adults they are
more vulnerable to psychiatric disorders when they experience stress or
trauma. Their offspring, whose own gametes are developing while in the
womb, might transmit these modifications to their own grandchildren,
refracting the experience of trauma across generations (fig. 1.1).[1]

   Poet-novelist Elizabeth Rosner, a daughter of Holocaust survivors,
wove Yehuda's findings into an extended reflection on intergenerational
trauma in her 2017 book *Survivor Cafe: The Legacy of Trauma and the
Labyrinth of Memory*. Research on how the fetal environment programs
our genomes, Rosner suggested, "is bringing us empirical proof of a
legacy we have already known in our bones, our dreams, and our ter-

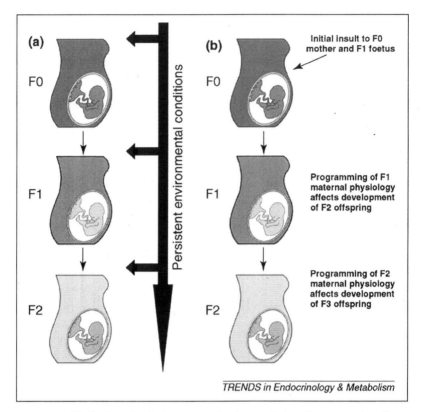

FIGURE 1.1 Mechanisms for the intergenerational transmission of programming effects. (a) Persistence of an adverse external environment can result in the reproduction of the phenotype in multiple generations. (b) The induction of programmed effects in the F1 offspring following in utero exposure (e.g. programmed changes in maternal physiology or size) results in programmed effects on the developing F2 fetus and so on. From Drake and Liu, "Intergenerational Transmission of Programmed Effects: Public Health Consequences." By permission of Elsevier.

rors. . . . Which is to say, my generation's DNA carries the *expression* of our parents' trauma, and the trauma of our grandparents too. Our own biochemistry and neurology have been affected by what they endured."[2]

As the matrilineal granddaughter of a Holocaust survivor just beginning my own family, I could not help but be curious about these claims. As a historian and philosopher of science who specializes in gender, genetics, and the social dimensions of scientific knowledge, the implications, both scientific and cultural, of the idea that a woman's health, behavior, and milieu can have intergenerational effects proved equally irresistible.

THE RISE OF FETAL ORIGINS SCIENCE

Soon, I realized that Yehuda's research was part of an efflorescence, in recent years, of scientific interest in the long-term effects of the intra-uterine environment. Beyond psychiatric disorders, scientists in a wide-ranging field of research on the fetal origins of health and life outcomes are searching for links between maternal factors such as diet, stress, and environmental exposures and offspring outcomes such as obesity, heart disease, autism, asthma, sexual orientation, attention deficit hyperactivity disorder (ADHD), and intelligence.[3]

While scientific interest in the imprint left by the womb has a long history, the proximate foundations of the modern-day field of fetal origins science—sometimes called Developmental Origins of Health and Disease, or DOHaD—trace to the late 1980s. Using historical health records from the poorest areas of England, in 1989 British epidemiologist David Barker demonstrated that communities with the highest infant mortality rates in the 1930s had the highest heart disease mortality rates 50 years later. Barker argued that the risk of heart disease in these populations was linked to development in the womb and hence could be seen as an outcome of the poor health status of their mothers, not only during the pregnancy itself, but throughout their own lives.[4] At the time, Barker's claims encountered disbelief and resistance. But today, three decades later, this hypothesis drives multiple lines of inquiry at the intersection of developmental biology, teratology, nutrition science, environmental science, and lifecourse epidemiology.[5] As of 2014, this research had pro-liferated considerably: more than 130,000 papers on fetal programming of disease were available in the biomedical research database PubMed.[6]

Compared to speculations on maternal intrauterine effects in previous eras, today's maternal-fetal effects science benefits from a greatly expanded body of epidemiological data on the prenatal period. Beginning in the late 1980s, researchers initiated large-scale prospective cohort studies tracking mothers and their offspring from pregnancy onwards. Interest in prenatal influences rose alongside two developments in the 1980s and 1990s. The first was the dramatic expansion of global public health investment in maternal and infant health. Policy makers, economists, and global development experts were increasingly highlighting the importance of the very early years for an individual's long-term health and economic well-being. Improving outcomes at birth became widely seen as a central site of intervention for advancing economic development in the world's poorest regions.[7] Contemporary with this was a dramatic upswing in public and private investment in the biosciences associated with the Human Genome Project. Powerful new prenatal genetic testing technologies introduced the prospect of predicting health risks from the earliest stages of fetal development. Researchers argued for the need to pair the study of these genetic vulnerabilities with research on prenatal environmental effects, not only to better understand the interaction between the two, but also to advance a parallel knowledge base on the links between early developmental exposures and later patterns of health and disease.

The Avon Longitudinal Study of Parents and Children (ALSPAC) at Bristol University in the United Kingdom, emblemizes these developments. The study began in 1991, enrolling 13,761 pregnant women. Today it is one of the largest studies attempting to track the long-term effects of prenatal exposures. ALSPAC has collected copious biological samples from its children and mothers, including maternal blood and urine during pregnancy and cord blood and placental tissues at birth. To date, the study has amassed data on children at 68 time points, including 9 clinical assessments. Even the children's baby teeth reside in the study's databank, collected by ALSPAC's appointed "tooth fairy." ALSPAC data have yielded more than 2,000 scientific publications, including findings on risk factors for obesity, eczema, and asthma.[8]

In the mid-1990s, joining the excitement generated by the major genome sequencing projects, ALSPAC collected DNA from 11,000 children and 10,000 mothers in the study. Now, ALSPAC is introducing epigenetic methods. Epigenetics refers to molecular changes in the non-DNA regulatory apparatus of the genome. Epigenetic markers that help determine whether a particular site on the genome is active or silent can change in response to environmental stimuli. If these changes can be shown to be able to be induced by the intrauterine environment, to remain stable over time once induced, and to have functional implications for human health and biology, epigenetics may offer a causal mechanism explaining the long-term effects of maternal pregnancy effects posited in DOHaD (pronounced "dough-had") studies.

But DOHaD science is not just mountains of data correlating prenatal exposures with later outcomes. Today's research on maternal intrauterine effects combines this new trove of data with a set of guiding assumptions that serve as a conceptual framework for interpreting these data and launching research hypotheses.

MATERNAL EFFECTS AND THE BIOSOCIAL BODY

DOHaD researchers believe that human developmental plasticity is greatest during the critical period of intrauterine growth, and that prenatal cues from the maternal environment can permanently program the developing fetus in ways that alter physiological functioning as an adult. One hypothesis is that the sensitivity of human fetuses to their maternal environment is an evolutionary adaptation, allowing fetuses to attune themselves to their expected postnatal environment. But if metabolic and stress signals from the mother's body do not match the actual world the fetus encounters—as in the case of a nutritionally deprived fetus entering a calorie-rich American dietary landscape—illness results. DOHaD founder David Barker influentially and provocatively appealed to this hypothesis to argue that many so-called Western diseases of affluence, such as breast cancer and heart disease, were driven by the rapid shift from the relative deprivation of early twentieth-century lifestyles to the hygienically and nutritionally transformed environment of the late twentieth century. An important prediction of this hypothesis is that because aspects of the intrauterine environment that a woman provides are themselves set by her own development in the womb, a "mismatch" between fetal programming and postnatal environment may "take several generations to disappear," as Barker has claimed.[9]

Many DOHaD researchers believe that, in this way, intrauterine effects help explain how persistent social inequalities become embodied and pass across generations. University College London pediatrician and child nutrition expert Jonathan Wells, author of *The Metabolic Ghetto* (2016), argues that rising rates of metabolic disorders and obesity are bodily manifestations of the intergenerational transmission of health inequalities. Wells believes that features of the mother's social and environmental context during her own development—including social class—are, in a sense, transmitted to the growing fetus, conditioning it for a life of inequality even before birth. "If pregnancy is a niche occupied by the fetus," Wells has written, "then economic marginalization over generations can transform that niche into a physiological ghetto where the phenotypic consequences are long-term and liable to reproduction in future generations."[10]

Similarly, Northwestern University anthropologist Chris Kuzawa hypothesizes that maternal hormones and nutrients provide the fetus "access to a cue that is predictive of its future nutritional environment."[11] He suggests that maternal signals to the fetus function "inertially" to prevent changes in the offspring that are too great and too rapid in a single generation. "The flow of nutrients reaching the fetus," Kuzawa believes, "provides an integrated signal of nutrition as experienced by recent matrilineal ancestors, which effectively limits the responsiveness to short-term ecologic fluctuations during any given pregnancy."[12] Kuzawa likens this to a "crystal ball" that "allows the fetus to predict the future by seeing the past, as integrated by the soma of the matriline."[13] Problems can arise, however, when this fetal environment for developmental modification and "fine tuning" is either "impaired" or mismatched with current environmental conditions.[14]

Applying this conceptual framework, Kuzawa and Elizabeth Sweet argued in a 2009 article that maternal effects may help explain persistent racial disparities in rates of cardiovascular disease in the United States.[15] Historically, African American women have experienced high rates of stress during pregnancy, in part due to experiences of slavery and its legacies of racism. This has contributed to lower birth weights, a predictor of later cardiovascular disease risk for offspring. Since a woman's own birth weight predicts her child's, maternal pregnancy effects provide a mechanism for the persistence of high cardiovascular disease rates across generations even after the diminishment of continued psychosocial or nutritional stressors. In this biosocial explanation of health disparities invoking maternal effects, racial differences are understood as social in origin, but mediated and transmitted by biological processes of early growth and development.

The science of "maternal effects"—defined by evolutionary geneticists Jason Wolf and Michael Wade as "the causal influence of the maternal genotype or phenotype on the offspring phenotype . . . generally through the maternally provided 'environment'"—offers a picture of heredity different than the one we learned in high school genetics.[16] It suggests that mothers endow the fetus with more than just DNA. As they develop, infants are programmed by maternal factors that influence the environment in which they grow. To many, research on maternal intrauterine effects shows just how profoundly our bodies are mediated by our environments, starting at conception. Applied to questions of persistent patterns of social inequality and to the phenomenon of intergenerational inheritance of trauma in human populations, maternal effects science offers a potentially powerful approach to understanding how our bodies are at once biological and social.

I am of two minds about such claims. Intuitively, I believe that bodies register their social and physical environments in ways subtle and profound, that health is a matter not just of biochemistry but also of social chemistry, and that the maternal-fetal relation is a powerful and mysterious one. Today, maternal effects science is part of a broader and, from my perspective, welcome turn away from gene-centric models of the determinants of health and human biological variation. The science of maternal effects suggests greater enigma and complexity in heredity than a simplistic genetic story will tell.[17]

But as intellectually exciting and socially important as it is to appreciate bodies and biologies as shaped by their environments, the present intensive focus on a narrow window of human development—gestation—and on a particular class of bodies—those presenting as women of reproductive age—requires scrutiny. Situating the intrauterine environment as a critically determinative one for a range of life outcomes articulates social alarm through gestational reproductive bodies in ways that carry real implications for restrictions on reproductive autonomy. Moreover, even if intrauterine effects exist, an exaggerated focus on the mother as the bearer of reproductive risk may misdirect resources from other, more important contributors to health and life outcomes.

Maternal effects science is also an area where the intellectual excitement runs ahead of actual empirical findings. Despite—or perhaps due to—the fact that today, scientists have more precise tools than ever for studying the biochemical relation between a mother and fetus, and access to massive, multidimensional sets of human biological and social

data across the life course, the science of maternal effects involves connecting unstable biological markers to small effects. In maternal effects
research, the effects under examination are often what I have come to
term "cryptic": they are small, vary depending on ecosocial context, and
frequently occur at a great temporal distance from the initial exposure,
making causality challenging to establish.

CRYPTIC CAUSALITY

The crypticity of the effects studied in today's fetal programming science is particularly brought into relief when compared with an earlier
class of studies of birth anomalies induced by prenatal exposures. Take
the case of folic acid supplementation, a public health intervention that
radically reduced birth anomaly rates around the globe within just a few
decades of its introduction. As early as 1965, researchers hypothesized
a relationship between dietary deficiencies in folate, a B vitamin, and
severely disabling neural tube defects, such as anencephaly (in which
large parts of the brain and skull are missing) and spina bifida (a class of
congenital disorders involving the spine and spinal cord formation). But
the hypothesis went untested until the late 1980s, when a large randomized control trial in England found that folate supplementation reduced
the rate of neural tube defects in offspring of women with a previous
pregnancy carrying this anomaly by an astonishing 70 percent.[18] Following the study's publication in 1991, public health agencies rushed into
action. In the US, the Centers for Disease Control and Prevention recom-
mended that women with a history of neural tube defects in pregnancy
take daily folic acid supplements. Soon after, the recommendation was
expanded to all women of childbearing age.[19] Starting in 1998, the US
required folic acid fortification of enriched cereal grain products. Combined, these interventions resulted in decreases of up to one-third in the
overall prevalence of neural tube defects. A 2011 review in the journal
Nutrients described folic acid fortification as "one of the most successful
public health initiatives in the past 50 to 75 years." A recent report even
suggested that with the saturation of fortification and supplementation
efforts, there is probably no intervention that could further reduce neural
tube defect rates—they are now as low as humanly possible.[20]

Like neural tube defects, blindness caused by rubella, limb shortening produced by thalidomide exposure, and iodine-deficiency dwarfism are outcomes immediately visible at birth. In these cases, teratogenic environmental toxins and nutritional deficiencies can be chemically specified and experimentally isolated in animal models. The effect sizes of the exposure are large; the consequences severe. The relationship between exposure and outcome is independent of ecology and is predictive across diverse human populations.

In contrast, the causal space explored by maternal effects science today is far more complex. Maternal effects research in human populations correlates the maternal pregnancy environment to offspring health outcomes. Compared to teratological research on birth anomalies, the outcomes reported in maternal-fetal effects science are not rare malformations immediately visible at birth. Instead, they are small changes in risk factors contributing to common chronic diseases—slightly higher body fat, for example—that might emerge several decades into life. Today, maternal effects researchers are not primarily studying the causes and later effects of atypical birth outcomes, but the long-term implications of small variations among apparently healthy, average births. The exposures of interest are not generally discrete chemical nutrients or teratogens, but diverse, confounded, and often unspecified stressors registered by the fetus while in the maternal milieu. The strength of the relationship between exposures and outcomes is not a biologically fixed constant, but varies depending on ecological factors in a population.[21]

In fact, the founders of the modern science of teratology—that is, the study of early developmental anomalies—did envision a wider possible remit for the field, one that would open investigation to the subtle effects of a wider range of small exposures in the prenatal period. The Society for Teratology's first president, the celebrated Polish-born American pediatrician Josef Warkany, opined as early as 1948 that there must "exist some borderline deficiencies" that lead to less starkly evident disabilities, but which nonetheless carry subtle long-term effects.[22] Just because "the young of deficient mothers may appear externally normal," Warkany suggested, "that does not mean they *are* normal."[23] Warkany recognized, however, that such "borderline" factors were not yet accessible to science. Warkany focused his own research on prenatal factors contributing to highly disabling developmental conditions visible at birth. In 1945, he discovered the link between prenatal iodine deficiency and a form of dwarfism. Later, he helped build the Cleveland Clinic into a leading international center for the care of children with birth anomalies. The study of "microenvironmental factors" at the maternal-fetal interface, he wrote in 1961, "will have to be left to a future science, a kind of micro-obstetrics, to elucidate the variables within an amniotic cavity."[24] The twenty-first-century field of DOHaD may be the "future science" Warkany anticipated.

Comparing teratological studies of birth anomalies and DOHaD research on prenatal influences highlights two distinguishing features of this more recent field of inquiry—the small size of the effects under study and the attenuated nature of the causal explanations offered. To embrace claims about the long reach of life in the womb, whether as a way of articulating intergenerational trauma at the level of the body, or as a basis of public health prescriptions to pregnant women, is to commit to a certain style of claims-making—a bolder, more permissive style of building hypotheses and making causal inferences than that which historically characterized the field of empirical teratology. Throughout the book, I use the term "cryptic" to describe these features of the maternal effects knowledge landscape.

The stakes of achieving a clear understanding of what is and can be known about causal maternal intrauterine effects are high, particularly when popularizations present the science of maternal-fetal epigenetic programming in simplified and alarmist terms. Nascent findings linking small risks with maternal behaviors are rapidly translated into everyday advice to pregnant people. The breathless reporting of DOHaD science— "Mother's Diet during Pregnancy Alters Baby's DNA" (BBC), "Grandma's Experiences Leave a Mark on Your Genes" (*Discover*), "Pregnant 9/11 Survivors Transmitted Trauma to Their Children" (*Guardian*)[25]—rarely offers sufficient context to judge the strength of the scientific claims, the relative size of the risks involved, the plausibility of extrapolation of findings from non-human models to human systems, or the likelihood of interacting confounding factors.[26]

The vexing matter of establishing causality when effects are cryptic, I argue, is especially treacherous in the case of maternal effects research because women's reproductive bodies, historically represented both as vectors of social contagion and as mysterious and unyielding objects of scientific investigation, are at the center of contention. But while the stakes may be particularly stark in the case of maternal effects science, the phenomenon of cryptic causality is not unique to this field. In today's data-rich world of biomedicine, with its rising tolerance for small effect sizes and low replication rates, the question of how and what we can scientifically know is being fundamentally transformed.[27] In this way, a close look at cryptic causality in maternal effects science and at the history of scientific debates surrounding how to reason about intrauterine effects in human populations offers insights into knowledge practices that increasingly characterize the twenty-first-century postgenomic life sciences.[28]

BEGIN BEFORE BIRTH

The intense research and media interest in epigenetics and the developmental origins of health and disease has today elevated to dizzying new heights a focus on women's reproductive bodies as critical drivers of costly social ills. The website BeginBeforeBirth.org, designed by researchers from the Institute of Reproductive and Developmental Biology at Imperial College London with funding from the Wellcome Trust (the leading biomedical research funder in the UK), greets visitors with two introductory videos: "What happens in the womb can last a lifetime" and "Epigenetics." The caption warns that "how the mother feels during pregnancy . . . can have a lasting effect on the development of the child."

BeginBeforeBirth.org prominently features "Charlie's story," a narrative designed to illustrate what these researchers see as the social costs of a mother failing to "begin before birth." Accompanied by an image of a teenage boy whose face is shrouded by a hoodie, we learn that Charlie, "caught looting in the riots" (presumably in reference to the 2011 London youth riots), is "just leaving prison" for stealing and is "often aggressive." All of this, we are told, is because "his mother was very stressed while she was pregnant," which "could have caused him to be a difficult baby and hard to soothe. It could also have led him, as he got a little older, to show signs of ADHD and being hard to control." "Charlie wasn't born a criminal," intones the accompanying video, "but research suggests that his time in the womb and his early life could have made his behavior more likely."[29]

University of Pennsylvania psychiatrist Adrian Raine similarly highlights the role of maternal-fetal effects in criminality in his 2013 *The Anatomy of Violence: The Biological Roots of Crime.* In a section subtitled "The mother of all evil—maternal neglect and epigenetics," Raine suggests that the maternally provided environment "not only changes gene expression in the individual—it also has permanent effects that transmit to the next generation." Citing the example of serial killer Henry Lee Lucas, Raine implies that the murderer's behavior resulted from the compounding of abuse by his "alcoholic prostitute mother" with "deprivation that she likely experienced herself as a child." The abuse was passed on "not just environmentally, not just genetically, but likely epigenetically," which "turned off important genes in Lucas that normally inhibit violence—and turned on genes that promote it."[30]

These sorts of claims have broad implications for public policy beyond the question of criminality. In an editorial titled "At Risk from the Womb," *New York Times* columnist Nicolas Kristoff cited DOHaD research to suggest that "a stressful uterine environment may be a mechanism that allows poverty to replicate itself generation after generation." Furthermore, "kids facing stresses before birth appear to have lower education attainment, lower incomes and worse health throughout their lives." As a result, Kristoff argued, social interventions that focus on infants and children may be misdirected: "Even early childhood education may be a bit late as a way to break cycles of poverty." DOHaD pioneer David Barker has lent support to this idea, arguing that fetal origins research might contribute "to our understanding of cognitive function" and is "tremendously relevant in education," where, "If you don't get it right by year three, forget it, your chances to go to Harvard are blown away."[31]

Researchers often justify the focus on maternal pregnancy effects by arguing that maternal behaviors are the easiest point of intervention to improve outcomes. They hope that evidence of the molecular biological changes produced by maternal behavior will lead women of reproductive age to alter their behavior in order to optimize their pregnancies. As Dutch epidemiologist Tessa Roseboom, a leading researcher in the DOHaD field, has put it, "In general, women are especially receptive to advice about diet and lifestyle before and during pregnancy. This should be exploited to improve the health of future generations."[32] A 2013 National Public Radio report captured the same sentiment, noting that "epigenetic studies may help get the attention of pregnant women who would otherwise ignore recommendations about diet and behavior." The report quotes Johns Hopkins scientist Dani Fallin, who also self-identifies as a mother, saying, "If you see there is a detectable biological change because of exposure to drinking or because of exposure to smoking, that as a pregnant mom would convince me that, oh, it matters."[33]

This reasoning embraces a punishingly expansive conception of individual maternal responsibility for the developmental environment of the fetus.[34] Take, for instance, a computer game intended to help participants experience the impact on the fetus of a mother's failure to provide a good intrauterine environment, available at BeginBeforeBirth.org. Each time the player—the mother—makes a mistake in a word-matching game, the screen shows a tiny fetus's heart rate ticking upwards. The game, which the website describes as designed "to show how when stressed, anxious or depressed during pregnancy . . . the resultant changes in [a mother's] physiology can be transmitted to the fetus, impacting upon fetal development and subsequent childhood behavioral and emotional responses," is indeed a distressing experience.

The everyday mundanity of the stressor—making a small mistake in a computer game—suggests not only that the risks are everywhere but that any amount of stress constitutes a risk to the fetus. "The early emotional environment can lead to long-lasting epigenetic changes in the brain," the website states. But stress is a fact of life, and we cannot always control our stressors. This is particularly true for the plurality of humans around the globe who experience persistent food, work, and housing instability, among other major stressors. BeginBeforeBirth.org, however, places the responsibility entirely on the individual woman. Featuring an image of a woman sitting alone, with brow furrowed and head in hand, the site directs women to control their stress by "doing yoga or listening to music." Here, stress is understood exclusively as feelings possessed by an individual woman—the social and structural context is invisible and unmentioned.

PREGNANT WOMEN AS FETAL CONTAINERS

For many readers, this focus on monitoring and modifying women's behavior during pregnancy will immediately bring to mind an earlier period, in the 1980s, of frenzy over the maternal threat to fetal health.[35]

Following high-profile incidents of birth defects caused by in utero exposure to toxins, such as the 1961 link between the morning sickness drug thalidomide and limb-shortening disorders, researchers began to intensively study the effects of prenatal exposure to environmental toxins and substances such as tobacco, alcohol, and illegal drugs.[36] These studies demonstrated the harmful effects of cigarette smoking on birth weight[37] and led to the description, in 1973, of fetal alcohol syndrome, an impairing disorder of physical and mental development in children of heavy drinkers.[38]

These efforts were quickly distorted as they were popularized in panicked terms and drawn into 1980s reproductive politics. Doctors reported women asking permission for a sip of coffee while pregnant.[39] Fear, later shown to be baseless, arose over whether white collar pregnant working women were exposing their fetuses to radiation by sitting in front of computer monitors.[40] Alarmist public health campaigns, showing heavily pregnant mothers drinking while their almost fully developed fetus is pictured, curled in the womb, sucking on a tiny beer bottle, exaggerated the effects of maternal behaviors during pregnancy by implying that the fetus directly experiences whatever the mother imbibes.[41]

Based on these distortions and claiming to speak in the interest of the fetus, policy makers enacted punitive measures on women with poor birth outcomes. In the US, hundreds of women were imprisoned, lost custody of their children, or were denied social welfare benefits on suspicion of drug use while pregnant. Anti-abortion advocates exploited nascent scientific findings about the effects of the prenatal environment to extend legal frameworks for child endangerment to the fetal period. By 1986, 21 states had passed or were considering instituting such criminal sanctions. A 1992 study documented 167 women imprisoned under these policies, 70 percent of whom were poor women of color.[42]

Imprisonment represents only the extreme end of a spectrum of autonomy-limiting efforts to sanction, proscribe, shame, or stigmatize pregnant women's lifestyle choices. The *Yale Law Review* reported in 1990 that some of the largest US companies explicitly refused to hire women of reproductive age in "settings that would be potentially hazardous to reproduction."[43] A survey of heads of fellowship programs in maternal-fetal medicine revealed that 46 percent "thought that women who endanger a fetus by refusing medical advice should be held against their will 'so that compliance could be assured,'" and 26 percent advocated state surveillance of women in the third trimester who stay outside of the hospital system.[44]

In his classic 1986 article, "Pregnant Women as Fetal Containers," bioethicist George Annas poignantly impugned these beliefs, citing the words of the lead character of Margaret Atwood's *The Handmaid's Tale*—a 1980s book itself inspired by the decade's increasing disregard for pregnant women's autonomy—"We are two-legged wombs, that's all; sacred vessels, ambulatory chalices." Annas pointed out that with the many claims flying around in the scientific literature about potential harms to a growing fetus, almost anything could substantiate a claim of "fetal neglect." "Does it mean that the pregnant woman must, in effect, live for her fetus? That she must legally 'stay off her feet' if walking or working might induce contractions? That she commits a crime if she does not eat only health foods; smokes or drinks alcohol; takes any drugs (legal or illegal); has intercourse with her husband?"[45]

The most dramatic example of a rush to judgment on the risks posed to fetuses by their mothers is the case of the so-called crack baby. In the 1980s, urban, impoverished African American communities experienced the devastating effects of crack cocaine, a cheap, highly addictive, and deadly drug. A 1985 scientific report suggesting that exposure during pregnancy causes developmental deficits in children led to a massive media amplification of the "crack baby," usually pictured as an undersized infant, born addicted to crack, and lesioned from the start.[46] Media reporting implied, as legal scholar and sociologist of science Dorothy Roberts has written, that "permanently damaged and abandoned by their mothers, [crack babies] would require costly hospital care, inundate the foster care system, overwhelm the public schools with special needs, and ultimately prey on the rest of society as criminals and welfare dependents."[47]

Women lost their children and faced criminal punishment, but little was done to offer addiction treatment that would allow them to keep their children with them, address the sources of violence, marginalization, and abuse in their lives that contributed to their addiction, or provide accurate and actionable information about changes individuals could implement that would make a difference. Instead, echoing longstanding anxieties about black women's reproduction as contributing to social pathology, the focus was on black women producing "transgenerational pathology," procreating a "bio-underclass" that is a "menace to society" through their "deviant lifestyle."[48] By the mid-1990s, follow-up research demonstrated that the alarm had not been justified—there are in fact no measurable long-term effects of crack cocaine exposure.[49]

As the case of the "crack baby" underscores, historically, there is frequently a vast disconnect between the degree of panic over prenatal risk factors and the bigger picture of factors contributing to poor health outcomes among infants. The sorts of lifestyle and behavioral risk factors that whipped public health officials into a frenzy in the 1980s pale in comparison to the legal, economic, and social-structural challenges standing in the way of healthy pregnancies and resilient childhoods. As essayist Katha Pollitt insightfully pointed out, "It would be pleasant to report that the aura of crisis . . . was part of a massive national campaign to help women have healthy, wanted pregnancies and healthy babies," but the "policies that have underwritten maternal and infant health in most of the industrialized West since World War II—a national health service, paid maternity leave, direct payments to mothers, government-funded day care, home health visitors for new mothers, welfare payments that reflect the cost of living—are still regarded in the United States by even the most liberal as hopeless causes." She continued, "It is an illusion to think that by 'protecting' the fetus from its mother's behavior we have insured [sic] a healthy birth, a healthy infancy, or a healthy childhood, and that the only insurmountable obstacle for crack babies is prenatal exposure to crack."[50] In short, an overfocus on pregnant women's individual behaviors can detract from the wider social context of her actions and from the responsibilities of both parents and the wider community to provision a safe and healthy environment for child development.

INTENSIVE MOTHERING IN THE POSTGENOMIC AGE

Today, the threats to women's reproductive autonomy posed by scientific claims about risks to the fetus are no less present—and in some ways are even more insidious—than they were in the 1980s. At that time, the science of maternal-fetal effects focused on discrete toxins ingested by the mother during pregnancy and passed across the placenta to the fetus. Today's postgenomic science of maternal effects, however, greatly expands the varieties of risky exposures. DOHaD scientists, concerned with the implications of the overall constitutional health of the mother for her offspring, believe that the total health of girls beginning when they themselves are in the womb, throughout childhood, and, during reproductive years, in the "preconception period," can contribute to the maternal imprint received by a fetus in the womb.[51] The exposures of concern are not limited to toxins; they include any "change in the overall endocrine or metabolic milieu, or both, of the fetus."[52] As Peter Gluckman and Mark Hanson write, fetal programming research goes beyond the "obvious point of avoiding toxins such as cigarettes and drugs of abuse." The focus is on "a more subtle level" at which "the fetus is constantly responding to signals from its mother."[53] The effects under study are not—as in the case of alcohol use—birth anomalies or long-term disability, but tiny alterations in bodily variables, such as an additional centimeter or two of abdominal fat at age 60. And, in DOHaD research, the implications of intrauterine effects are not limited to a mother's offspring but are hypothesized to possibly refract through her descendants via epigenetic inheritance.

Furthermore, since the 1980s, women themselves have much more thoroughly internalized the idea that it is their responsibility to look to science for guidelines on how to adapt their lifestyles to optimize their offspring.[54] Maternal-fetal technologies, such as sonograms and prenatal testing, as well as pervasive public health messaging about fetal risk factors have, as philosopher Quill Kukla writes, transformed pregnant women's "own relations to the insides of their bodies."[55] We now "treat pregnancy as something that is and ought to be *designed*, and designed in accordance with public, social standards," writes Kukla.[56] With thousands of papers published each year correlating pregnancy exposures and later outcomes, women are subject to a continual barrage of statistics reporting links between everyday behaviors—stress, exercise, hot tub use, consumption of sugar—and outcomes for their future children, such as ADHD or obesity. It has now become, Kukla writes, "part of what makes us count as a *conscientious* expectant or new mother that we consume public pregnancy information such as guides and websites; carefully document numbers of kicks, urinations, etc.; report these results to the proper authorities; have doctors measure our weight, blood sugar level, and fundal height at prescribed intervals; and so forth."[57] Because of this cultural conditioning, DOHaD research is poised to be received more uncritically, and integrated and internalized more rapidly into women's lifestyle choices, than research in any previous historical era.[58]

The force of social pressures to maximize fetal outcomes can be seen in the introduction of non–evidence-based policies putatively to safeguard the fetus but that come at the risk of women's well-being or autonomy. For example, on the theory that gestational diabetes is linked to future offspring obesity risk, the International Association of Diabetes and Pregnancy Study Group recently updated its recommendations for gestational diabetes diagnostic criteria by lowering the threshold for diagnosis.[59] More women will now be diagnosed with gestational diabetes, which requires often dramatic revision of household diet, but there is little evidence linking this to a benefit for future offspring's risk of obesity.[60] Relatedly, women are again being weighed more frequently throughout pregnancy, a practice that had begun to fall by the wayside due to lack of evidence for benefits combined with a possible link to maternal anxiety and eating disorders.[61]

In another instance of such non–evidence-based policy, in 2016, the US Centers for Disease Control and Prevention issued an advisory urging all sexually active women of reproductive age to abstain from alcohol use, even if they were not intending to become pregnant.[62] Although heavy drinking during pregnancy is statistically linked to developmental deficits in offspring, no evidence supports a requirement for absolute abstention prior to or at conception.[63] For some, it may be easy to dismiss the notion that alcohol use is a component of human autonomy. But given the pleasurable (for many) place of moderate alcohol consumption in the everyday byways of a typical young American woman, such a proscription represents a significant impingement on women's autonomy, all in the name of protecting a hypothetical unplanned fertilized embryo from a substance for which the unsafe level of exposure is not scientifically established.

This emphasis on fetal well-being at the expense of the mother is increasingly enshrined in the law.[64] Driven by anti-abortion activists who argue that the fetus should be granted the same rights as any individual, in the greatest roll-back of United States women's reproductive autonomy since the pre-*Roe* era, some 38 out of 50 US states now extend legal structures originally designed to protect children against abuse and endangerment to the fetal period, according to a 2018 seven-part series in the *New*

York Times that documented how fetal protection laws are insidiously used to surveil, punish, and restrict women.[65] Inevitably, scientific research that implies that the fetal period is one of greatest vulnerability, and that the mother conveys the total critical developmental environment for the child, is cited by advocates for these laws and contributes to the perceived need to establish legal protections for the fetus.[66]

This unrelenting and often uncritical focus on women's behaviors as the origin of health and societal woes is a reality within which new research on maternal intrauterine effects in humans is situated, regardless of scientists' intentions. The question of reproductive rights, responsibilities, and justice therefore forms one important dimension of this historical, philosophical, and social study of the science of maternal intrauterine effects.

And here, a note on terminology is in order: Coextensive with the language employed in the historical moments and scientific fields I engage in this book, I will often use the terms "maternal" and "women" when discussing maternal effects science. This terminology is both imprecise and non-inclusive. Not all mothers have a uterus, nor do all women have a uterus. Not all gestational parents identify as mothers, and not all mothers identify as women. And, of course, not all women are mothers. Maternal-fetal effects science is about gestational exposures, but it carries implications, albeit ones that vary depending on social status, for anyone who is hailed as having a uterus or being a mother, regardless of reproductive anatomy or gender, sexual, or parental identity.

A HISTORY OF MATERNAL EFFECTS SCIENCE

This book examines the intertwined social and scientific dimensions of scientific speculations about the long reach of the maternal intrauterine imprint. Presenting a history of maternal effects science from the advent of the genetic age to today, *The Maternal Imprint* analyzes the entanglement between, on the one hand, conceptual and empirical debates over how to study maternal effects given their persistent crypticity, and, on the other, the implications of maternal intrauterine effects science for women's well-being and autonomy. The episodes I have selected track these two threads, featuring the premier scientific developments in debates over the existence and significance of maternal effects in heredity and development, and, at the same time, what these scientific claims were taken to imply about sex equality and parental reproductive responsibilities. Maternal effects science is a multidisciplinary field with implications across many arenas and which draws upon diverse genealogies across the globe. Naturally, the account presented here is a partial one, a Kandinsky rather than a Seurat. The result places diverse episodes in proximity and tension with each other, producing a conceptual map of twentieth-century dialogues about the pregnant body in intergenerational time.

One of the greatest provocations of maternal effects science is that it contravenes a founding premise of the modern theory of genetic inheritance: that males and females contribute equally to the next generation. *The Maternal Imprint* begins by reconstructing the principle of sex equality in heredity as it was first articulated in the late nineteenth century. August Weismann's theory that the egg and sperm are isolated from the ravages endured by other bodily cells, and that the female and male gametes contribute equally to the next generation, enacted a radical break from previous reigning ideas about heredity. Folk belief as well as leading scientists and physicians had long held that maternal and paternal contributions to the next generation were different. Among these theories was the longstanding notion that a mother's countenance, ambient environment, diet, and fleeting impressions during pregnancy could imprint on her developing fetus, marking it for life and contributing to its future character. Weismann rebuked these theories, presenting the refreshing but polarizing idea that each generation begins anew and that, beyond fulfilling basic nutritional needs, the maternal environment carries no particular consequence.

Theories that a mother's fleeting thoughts, sensations, and visual impressions could imprint upon a growing fetus are often claimed to have largely receded from respectable discourse by the end of the eighteenth century. But the upswing in maternal impressions theories of the role of the intrauterine environment in heredity and development in the late nineteenth and early twentieth century shows that this is not true. During this period in the United States, for example, "prenatal culturists" combined eugenic and progressive-era ideas about social uplift through bodily conditioning with new theories about the critical importance of very early experience in brain development to elevate the importance of maternal behaviors during gestation. Intriguingly, they proposed that the importance of prenatal culture should compel recognition of women's freedoms, including a right to an education, so that she can maximally prepare her growing fetus for responsible citizenship. Rejecting Weismann's posit that genetic material passes from generation to generation untouched by its milieu, prenatal culturists positioned the conditioning of the plastic fetus as a mode of controlling one's biological fate and overcoming a negative hereditary endowment. Against Weismann's dismissal of maternal impressions as superstition that is inconsistent with the laws of heredity, the prenatal culturists offered a different stance toward scientific knowledge. They appealed to everyday experience, common sense, and also the likelihood that some realities may escape scientific probes and measures.

By 1920, however, prenatal culture theories were firmly rejected by the mainstream scientific and medical establishment, completing the rupture from previous beliefs about the role of the womb in heredity initiated by Weismann. Biological scientists and medical doctors now dismissed any notion that the mother—except in cases of extreme deprivation or injury—could alter her offspring's traits. Consensus asserted that the fetus was walled off from a woman's body by the placenta and that a child's fate was set by a combination of its genes and its post-birth upbringing. Pregnancy advice manuals instructed women in the new science of genetic inheritance and warned them not to be swayed by fears of maternal impressions. In prenatal advice to prospective parents, scientific eugenicists emphasized, above all, the importance of the health and maintenance of the germ plasm, contained in each sperm and egg.

Because adherents of Weismann's germ plasm theory of heredity conceptualized prenatal health as the proper care and hygiene of the sperm and egg, interest in parental reproductive responsibility and early developmental exposures now focused on both parents and on the period prior to conception. At the time, prevailing ideas held that environmental risks to the human gene pool were most likely to be introduced by men, with their riskier lifestyles, occupations, and consumption patterns. Thus, in contrast to the prenatal culturists and to DOHaD scientists today, scientists in the 1920s and 1930s focused their prenatal advice far more on the preconception period, and on men, than on intrauterine exposures, and women. Men were exhorted to protect their sperm by abstaining from alcohol, undergoing regular syphilis testing and treatment, and maintaining their bodies in overall good health. Women, on the other hand, were instructed to evaluate potential partners by the avidness of their healthy-sperm-producing lifestyle. Reflecting on this period throws into dramatic relief the peculiarity of today's view that a focus on maternal pregnancy effects in producing offspring outcomes is just "common sense."

How did we get from there to here? During the same decades that human geneticists and medical doctors were waving away the possibility of maternal intrauterine effects on heredity and development, anomalies to Mendelian patterns of inheritance began to accumulate in the scientific literature on plant and animal genetics. For some traits, it appeared that maternal phenotype was always determinative, regardless of paternal contribution. Scientists wrestled with how to interpret these findings, eventually agreeing on the ambiguous term "maternal effects" to refer to a wide class of observations of maternal influence on the next genera-tion, in the absence of genetic transmission. In applied contexts, such as agricultural breeding of plants and livestock for desired qualities, the mass production of mice for use in scientific research, and the develop-ment of in vitro fertilization technologies for the treatment of human infertility, maternal intrauterine effects became not merely a matter of an intriguing anomaly to Mendelian theory, but a force to be reckoned with in achieving human aims. In the 1940s and 1950s, new techniques in re-productive biology allowed maternal effects to be studied experimentally in nonhuman placental mammals. These experiments established that maternal effects can alter heritability estimates, for example by changing rates of fetal growth. Evidence that the intrauterine environment could matter, particularly for macrosomatic traits like body size, led some to wonder whether maternal effects might be at work in human health.

In the 1950s and 1960s, these speculations were taken up in new research on variation in birth weight among human populations. On aver-age, black Americans are born smaller than white Americans. The origins of this disparity, and its implications for health and social inequities across lifespans, have long been a matter of debate among biologists, epi-demiologists, and social scientists. Birth weight began to be regularly col-lected as a demographic variable beginning in the 1950s, and researchers seized upon it as a rich biosocial measure of fetal well-being. In contrast to previous ideas that held that the racial disparity was genetic, scientists, drawing on maternal effects theories, forged the hypothesis that racial differences in birth weight reflect racial discrimination and deprivations, which can leave a mark even before birth. They also posited that the intrauterine environment might serve as a mechanism for transmitting these effects across generations. These maternal-fetal theories of racial variation in birth weight entered wider 1960s and 1970s debates over the causes of intransigent racial inequalities, including the controversy over genetic theories of racial gaps in IQ and educational attainment.

Debates over racial differences in birth weight from the 1950s to the 1970s raised the prospect of mining the prenatal period for yet-to-be-discovered drivers of health and social inequalities. Scientists theorized prenatal growth metrics as ripe material for formulating biosocial hypotheses about how inequality gets under the skin. In doing so, they linked maternal effects science to socially and politically powerful narratives and intellectual frameworks that conceptualize present-day human bodies as dialectically constructed by the past environments of their ancestors, as products of power structures and politics, and as vectors through which we can today change relations of power. Appreciating this entanglement of maternal effects science with the project of elucidating and ameliorating the biological dimensions of social inequalities is important to understanding the trajectory of the field from the 1960s up to the present day.

Today, scientists view birth weight as too crude a measure of fetal outcomes to support causal hypotheses about maternal pregnancy effects. Instead, maternal-fetal effects researchers have turned their attention to the new science of epigenetics. Epigenetics is a postgenomic science, exploding in the years since the conclusion of the major genome sequencing projects around 2010. It offers the prospect of understanding the genome as a dynamic, reactive system, integrating environmental inputs. It also provides a molecular mechanism for the transmission of environmental imprints across generations. All of this has led to great excitement about epigenetic science across many fields of the life and social sciences.

But there exist large methodological and conceptual obstacles to establishing epigenetic mechanisms as a stable, informative marker of maternal intrauterine exposures, and as a specific cause of later outcomes. Simply put, the notion that methylation levels represent a valid and generalizable measure of social processes and structures as experienced by diverse individuals and populations in the womb remains speculative.

Through a review of three major research streams in the field, I demonstrate how the current level of interest and excitement about maternal-fetal epigenetic programming outstrips what the science actually can and does show.

Fundamental questions about what claims are warranted in the arena of maternal effects science, given the crypticity of the causal processes and effects under study, are not just a recent development in the field. As *The Maternal Imprint* shows, scientists have grappled with what science can actually know about maternal effects since the days of Weismann. In each historical period, views about what can be scientifically studied, and indeed known, about human maternal effects are shaped by historically situated convictions about the importance, relative to other factors, of the maternally provided environment to human health and life outcomes.

Today's DOHaD research finds small effects that are significantly removed in time from the posited initial developmental exposure. These exposures are frequently patterned by socioeconomic status. Furthermore, paternal and maternal effects are rarely symmetrically examined in a study population. Nonetheless, unreplicated findings of small effect sizes and tenuous evidence for epigenetic mechanisms in this field of research are widely embraced as foundations for scientific hypotheses and rapidly translated into advice to women and pregnant people. One reason for this, I argue in the final chapter of *The Maternal Imprint*, is that social assumptions make some causal stories appear more logical and plausible than others. In maternal-fetal effects science, an enduring prior acceptance of widely shared social assumptions about where agency and responsibility lie for offspring outcomes works to elevate the intelligibility and seriousness with which cryptic scientific claims asserting that the ultimate responsibility for offspring outcomes lies with the mother are regarded.

The history of maternal effects science in the twentieth and twenty-first centuries that I relate in *The Maternal Imprint* might challenge some core stories that we tell about the history of the present. The standard narrative is that today's renewed interest in gene-environment interactions and intergenerational inheritance represents a corrective return, after a period of gene-centrism, to long-neglected and ostracized ideas emphasizing the importance of agency and the environment in directing adaptive biological systems. Placing instead the history of gendered reproductive bodies and the question of sexual difference at the center of our analytic axis, I argue in the epilogue to this book, casts a fresh light on narratives of continuity and rupture in our present genealogies of newly emerging progressive sciences such as epigenetics, postgenomics, and biosocial biology.[67] Fully acknowledging the persistently unresolved problematic of sexual difference in heredity within the twentieth- and twenty-first-century life sciences yields what, to some, may be surprising conclusions, such as the realization that Weismann's gene theory and a century of gene-centrism provided a toehold for a radical critique of enduring theories of sexual difference in heredity, development, and reproduction—and that epigenetics and the postgenomic biosocial sciences may unwittingly provide the material for the rearticulation of those theories.

But at a much more practical level, I wrote this book for people who have, or will have, babies, and those who care about them. I am inspired by smart critics, like the economist Emily Oster in her book *Expecting Better* and the scientist Amy Kiefer at the blog *Expecting Science*, who in recent years have applied a ferocious critical eye to everyday advice to pregnant people and new parents.[68] Upon investigation, these critics find that the advice exaggerates fetal risks and often lacks a credible evidential base. This book does not offer pregnancy advice, but it does join the call for scrutiny of scientific advice given to expecting parents in a different way, using the tools of history, philosophy, and gender studies of science to offer insight into how and why claims about the long reach of the womb are at once beguiling, challenging to validate, stubbornly persistent once launched, and beset by scientific controversy.

CHAPTER 2

Sex Equality in Heredity

"Neither gamete can be regarded as superior to the other . . . the living spark is not the exclusive property of either," wrote feminist philosopher Simone de Beauvoir in "The Data of Biology," the opening chapter of her iconic *The Second Sex*. "The nucleus of the egg is a center of vital activity exactly symmetrical to the nucleus of the sperm. . . . As a matter of fact, the embryo carries on the germ plasm of the father as well as that of the mother and transmits them together to its descendants under now male, now female form. It is, so to speak, an androgynous germ plasm, which outlives the male or female individuals that are its incarnations."[1] The sexless "germ plasm" to which Beauvoir appealed in her dismissal of biological warrants for political sex inequality was the progeny of a revolution—not a social one, per se, but a revolution in biology—that took place at the turn of the twentieth century, galvanized by German biologist August Weismann's theory of the continuity of the germ plasm.

Before chromosomes and genes, there was the "germ plasm." Developed in the 1880s, Weismann's germ plasm theory of heredity, which overturned prior theories holding that organisms can transmit acquired traits to offspring, comprises three propositions. First, that the hereditary material is a physical substance containing the instructions for the distinct traits characterizing an organism's ancestry. Weismann called this substance the "germ plasm." Second, that the germ plasm is passed from generation to generation through the nuclei of the "germ-cells"— the sperm and the egg. Third, that the germ cells are physiologically wholly distinct and sequestered from the bodily cells. While the body's cells differentiate during development and bear the marks, alterations, and vulnerabilities of the individual organism, the germ plasm remains intact and unaltered. This proposition is known as "the continuity of the germ plasm." Weismann's theory offered a concise physical and mechanistic account of heredity as the transmission of particles via the germ cells, irrespective of the life history of the animal. According to this theory, all hereditary transmission takes place at the moment of fertilization with the joining of equal amounts of nuclear material from the egg and the sperm.

"The physiological values of sperm and egg-cell are equal; they are as 1:1," wrote Weismann in 1880. "We can hardly ascribe to the body of the ovum a higher import than that of being the common nutritive basis for the two conjugating nuclei. . . . The germ-plasma in the male and female reproductive cells is identical."[2] This insight, Weismann later wrote, formed "the keystone for the whole structure" of what became the germ plasm theory of heredity, ultimately leading to a transformation in biological understanding of reproduction, evolution, and heredity that became the conceptual foundation of twentieth-century genetics.[3]

Weismann's germ plasm theory, in time, swept away the previous century's reigning Lamarckian theory of inheritance of acquired traits, which held that adaptive characteristics acquired by the parent during its lifetime were passed to its offspring. As many soon recognized, it also eviscerated one of the most durable pillars of the material foundations of human sex and gender roles: the asymmetric role of the human male and female in reproduction. With his concept of sex equality in heredity, Weismann challenged prevailing nineteenth-century theories of the differential contribution of the male and female to heredity and reproduction, including the theories, discussed below, of maternal impressions, spermic rejuvenescence, and male and female complementarity in hereditary transmission.

The germ plasm theory is most familiarly situated within the history of modern genetics. But approaching it from a different direction— the history of scientific theories of sex—unveils the underappreciated groundwork that Weismann's germ plasm theory supplied for an epic transformation in the theory of sexual reproduction. Mid-nineteenth-century accounts of heredity and sexual reproduction assumed that males and females played different roles in heredity and made qualitatively and quantitatively distinct hereditary contributions to offspring. Moreover, "heredity" was not universally assumed to be constituted solely by the transmission of material at the moment of conception. Rather, heredity encompassed the totality of characters shaped by the mother and father before birth. Operating at the interface between the natural and social order, these theories fueled intricate philosophies of what it is to be a male or a female, what social roles men and women should have, the responsibility of mothers and fathers for reproductive outcomes, and even the relative superiority of one sex or the other.

Tracing the development and impact of Weismann's provocative premise of sex equality in heredity presents a study in how fresh scientific ideas at the nexus of biology and culture can transform the terms of entire realms of human understanding and open the conditions of possibility for new political horizons. Like Beauvoir, many sensitive interpreters of gender and science have posited the arrival of Weismann's germ plasm theory and, following on its coattails, the chromosomal theory of heredity, as a transformative moment in the history of biological justifications for male and female sex inequality. At the close of her powerful classic treatment of the nineteenth-century sciences of male and female mental and physical inequality, *Sexual Science*, historian Cynthia Russett asserted that "evidence that the father and mother each contributed exactly half of the genetic endowment of their children of both sexes" clinched the downfall of famously misogynistic Victorian paradigms of the science of sex differences.[4]

Weismann's principle of sex equality in heredity, coupled with the subsequent rediscovery of Mendel's laws in 1900–1902, introduced a stark historical rupture in theories of heredity and reproduction—and, perhaps unwittingly, their trailing gender politics—at the advent of the twentieth century. What follows is not a study of Weismann's personal political beliefs about gender equality, however.[5] Rather, at our embarkation point, my interest is to reconstruct this historical rupture within the history of the sciences of heredity and sex—for one of the central stakes of the twentieth-century debates over heredity and maternal effects chronicled in this book is the hard-won and much-coveted idea of the equality of male and female contributions to heredity.

WEISMANN AND THE NEW BIOLOGY

Born in 1834, August Weismann began his career in the life sciences in the middle of the nineteenth century just as Charles Darwin's theory of evolution was transforming the study of biology. His major contributions to our understanding of the material basis of heredity, however, came after Darwin's death, in the late nineteenth century. Weismann, who was director of the Institute of Zoology in Freiburg, Germany, from 1870 to 1912, left a voluminous record of the evolution of his thinking.[6] In dozens of essays beginning in the early 1880s with "On Heredity," his inaugural lecture as Pro-Rector of the University of Freiburg, and culminating with the wide reception of his book-length text *The Germ-Plasm: A Theory of Heredity* in the 1890s, Weismann publicly grappled with evidence, sparred with opponents, and honed and revised his ideas.[7]

Eschewing technical language wherever possible, and interweaving historical and philosophical reflection, the essays make for a surprisingly pleasurable read. Weismann wrote with rhetorical craft and clever turns of logic, soon to be a lost art as the life sciences entered an intensive period of professionalization and specialization. On the translation of his *Essays* into English, his editor correctly foresaw that Weismann's work would "interest many who are not trained biologists, but who approach the subject from its philosophical or social aspects."[8] Weismann's theories of heredity were debated in leading periodicals of the day. In one notable exchange, Weismann and British sociologist Herbert Spencer clashed in the literary and current affairs journal the *Contemporary Review* over the plausibility of the inheritance of acquired traits, a back-and-forth that drew wide attention and commentary from the intelligentsia. This included socialists such as George Bernard Shaw and H. G. Wells, who vigorously debated the implications of Weismann's denial of the possibility of the transmission of acquired traits for the prospect of human political evolution toward a more egalitarian society.[9]

During the late nineteenth century, biology was still a young scholarly discipline. Scientists who in various ways studied the organic world were working to consolidate studies in zoology, animal science, physiology and anatomy, and biochemistry that had accumulated in the first half of the century into a body of coherent knowledge. Simultaneously, they were self-consciously building biology as a modern science, shunning the non-experimental observational sciences and cleaning house of what they saw as the relics of folk belief and the vitalistic sciences retained in received doctrine. In contrast to older methods of gross descriptive and comparative biology, and to the bulky tomes of biological observations bequeathed by their forebears, many German scientists of Weismann's generation sought a biology grounded in broad, unifying principles and in the experimental search for biochemical mechanisms at the level of the cell and the particle.

The most formidable zoological theory of the day, of course, was evolution by natural selection, first introduced by Darwin in 1859. The theory of natural selection asserted that all life evolved from a common ancestor, diversifying through the acquisition and hereditary transmission of small variations conferring adaptive advantage to individual organisms in the quest for survival and reproduction. In the late nineteenth century, however, significant elements of this theory remained contested even amongst its most avid adherents. Perhaps most worryingly, no one could explain how the action of natural selection operated on the body so as to transmit advantageous traits to future generations. Weismann was a convinced Darwinian—his preeminent biographer has even dubbed him the original "Neo-Darwinian"—keen to place Darwin's theory on the foundation of chemistry and physiology, requiring no metaphysical presuppositions or leaps of faith.[10] Heredity, he asserted, must ultimately reside in a "substance with a definite chemical, and above all, molecular constitution."[11]

Of the existing theories of the material basis of heredity, the worst offender, in Weismann's estimation, was pangenesis. The theory of pangenesis held that reproductive cells gather particles from all of the bodily tissues. These particles constitute the biological material required to replicate the organism. Through pangenesis, the ancestral traits of the living organism, as well as its acquired alterations, pass to its offspring. The theory's most famous booster was Weismann's guiding light, Darwin himself, who though recognizing its provisional and speculative nature, proposed it as a plausible schema for conceptualizing the production and preservation of biological variation at the level of reproductive cells.[12]

The theory was widely debated, with leading scientists such as Darwin's cousin Francis Galton inveighing against it and others coming to its defense. Weismann saw in the theory of pangenesis the ghosts of the old biology. Attempts to render such a theory in any way physiologically plausible, he charged, left scientists "compelled to suspend all known physical and physiological conceptions" and forced to accept "entirely gratuitous assumption(s)."[13] And here we find the stirrings of what is perhaps Weismann's least appreciated contribution—his philosophy of science.

It is a matter of some irony that Weismann became most widely known, beyond the pages of scientific journals, for a simple and vivid experiment. For several generations, he snipped the tails of white mice. He then bred the short-tailed mice with one another, but, alas, all of the offspring were born with long tails. The "acquired" mutation of a short tail never appeared in the offspring. Weismann's mouse-tail experiment, which debuted in his 1888 essay "The Supposed Transmission of Mutilations," was the rebuke to the theory of inheritance of acquired traits that became the popular trope by which his views found their widest reception. The stunt became the mnemonic by which a century of school children learned the folly of the nineteenth-century theory of inheritance of acquired traits.

But Weismann was principally a theorist, not an experimentalist. Rather than the reportage of empirical findings, Weismann aimed to introduce a theoretical framework that could synthesize observations about the physiological transmission of hereditary traits at the moment of fertilization with the theory of evolution by natural selection and the data of embryology. "The time in which men believed that science could be advanced by the mere collection of facts has long passed away," he urged. Science involves not the accumulation of "a vast number of miscellaneous facts," but the assemblage of "facts which, when grouped together in the light of a theory, will enable us to acquire a certain degree of insight into some natural phenomenon."[14]

"To go on investigating without the guidance of theories, is like attempting to walk in a thick mist without a track and without a compass," Weismann pled. Science, Weismann argued, was not the accumulation of observations as in a bricklayer constructing a building, but the construction of deeply penetrating insights, as in the drilling of a mine that opens into miles of productive yield. "Instead of comparing the progress of science to a building, I should prefer to compare it to a mining operation, undertaken in order to open up a freely branching lode."[15] Mining rather than building; voyaging with a compass rather than wandering blindly in the mist—with these vivid metaphors, Weismann defended an epistemologically sophisticated picture of science as theory-laden, revisable, and premised on the explanatory unity of the physical and organic worlds. Science is not "an edifice . . . solidly built by laying stone upon stone," from "the simple to the complex," but a "coherence" arrived at "analytically and inductively, proceeding from above downwards," he wrote.[16]

The opening premise of Weismann's germ plasm theory was the premise that "hereditary tendencies are as strong on the paternal as on the maternal side."[17] From this, Weismann predicted several necessary conditions of any biological account of the physical basis of heredity. First, the quantity of hereditary material should be equal in the male and female gametes. "Is it not possible," Weismann reasoned, "to draw a perfectly distinct and certain conclusion as to the relative quantity of the material basis of heredity, present in the germ cell of either parent, from the fact that the father and mother possess an equal or nearly equal share in heredity?"[18] He therefore surmised that "the amount of the substance which forms the basis of heredity is necessarily very small," since it must fit inside the tiny sperm as well as the much larger egg.[19]

Based on these suppositions, Weismann deduced that hereditary material in sperm and egg is the exact same sort of substance, in kind. It has no sex whatsoever. "To condense my argument into a sentence," he wrote, "we ought not to speak, as formerly, of the two conjugating nuclei of the germ cells as male and female, but as *paternal* and *maternal*; they are not opposed to each other, but are essentially alike, differing only as one individual differs from another of the same species."[20]

Starting in the early 1880s, Weismann set out to assemble, from supposition and suggestive but partial facts, a theory of the biochemical action of the egg and sperm in sexual reproduction. To summon a plausible material account of the transmission of hereditary materials at the moment of fertilization, Weismann turned to the rapidly developing but still sparse sciences of the physiology of sexual reproduction and embryology of the egg and the sperm. Weismann, who himself had been recently studying the development of egg and sperm in jellyfish but whose experimental work had increasingly gone by the wayside as he suffered the gradual loss of his vision, gleaned whatever supportive findings he could from the research of embryologists studying the reproductive biology of worms and sea creatures.

One experiment of particular interest came from the German embryologist Theodor Boveri. Working with sea urchins, Boveri removed a nucleus from a fertilized egg of the species *Sphaerechinus* and then replaced it with a nucleus from the *Echinus* sea urchin.[21] The result, according to Boveri, was offspring resembling *Echinus*. Nuclear transplantation experiments of this sort, Weismann argued, suggested that the (maternal) egg cytoplasm played no role, and that "the nuclear substance is the sole bearer of hereditary tendencies."[22]

The work of Berlin zoologist Oscar Hertwig also profoundly influenced the development of Weismann's theory of sex equality in heredity. In his earliest 1880s writings on the principle of sexual equality in heredity, Weismann drew on Hertwig's famous study of fertilization in sea urchins.[23] Hertwig observed the impregnation of the ovum by a sperm and fusion of their nuclei, forming a new single nucleus in the resulting zygote.[24] Weismann interpreted Hertwig's finding that sexual reproduction was accomplished by the conjugation of a single sperm with an egg and the fusion of their nuclei as clear evidence that the hereditary material resides in the nucleus and that the nuclei of the egg and sperm are identical in structure and function with respect to the mechanism of hereditary transmission. Later, in the 1890s, Weismann found encouragement in Hertwig's studies of egg and sperm formation. Hertwig and other workers had now shown that in both males and females, gametes are formed by a double division (today known as meiosis), dividing twice to form four gametes.[25] To Weismann, this provided additional evidence for the essential equivalence of egg and sperm nuclear material.

In retrospect, the meaning of these findings may seem glaringly obvious. Today, high school biology textbooks present the theory of the continuity of the germ plasm as the inevitable conclusion of the revelation of the workings of neatly aligned chromosome pairs during meiosis. But at the time, these findings represented new and still-contested science. The inheritance of acquired characters and the notion that males and females made distinct contributions to heredity and development was widely embraced. The evidence looked more like the broken pottery shards out of which an archaeologist puzzles the habits of a whole civilization. Indeed, other scientists looking at the same evidence had come to very different conclusions than Weismann.

MATERNAL AND PATERNAL CONTRIBUTIONS TO HEREDITY IN THE NINETEENTH CENTURY

By the mid-nineteenth century, scientists in the growing fields of embryology, cytology, and microbiology were in avid pursuit of the physiological mechanism of heredity and reproduction at the level of the union of the egg and sperm cells. Framing these investigations was the view, held in various forms for centuries, that male and female parents offer distinct and unequal elements to their offspring.

Maternal Impressions and the Inheritance of Acquired Traits

Maternal impressions theories—which hold that a pregnant woman's emotions and experiences can imprint on the fetus, leading to features such as birthmarks, deformities, and distinctive personality traits— constitute the most robust and longstanding doctrine concerning the mother's role in producing characters in offspring.[26] Diverse beliefs in the power of maternal emotional lability, visual encounters, and bodily condition to imprint on the fetus appear throughout the history of Western medicine. Common in premodern theological, natural-philosophical, and agricultural texts, tales of maternal impressions explaining birthmarks, rare talents, so-called monstrous births, and other anomalies convey an enduring conviction that a mother's thoughts, emotions, milieu, and habits impress upon the growing fetus, influencing its future physiognomy, character, and life prospects.[27]

To prevent monstrous births, French surgeon and lay teratologist Ambroise Paré warned in 1573 against intercourse on Sundays or religious holidays, too frequent intercourse, sexual excess in women, and sex during menstruation. Paré invoked "ardent and obstinate" maternal imagination as a prolific cause of deranged births: "the force of the imagination being joined with the conformational power, the softness of the embryo, ready like soft wax to receive any form."[28] He advised readers "how dangerous it is to disturb a pregnant woman, to show her or to remind her of some food which she cannot enjoy immediately, and indeed to show them animals, or even pictures of them, when they are deformed and monstrous."[29]

Maternal impressions theories, though diverse in mechanism and scope, were united in asserting a special contribution of the maternal-fetal relation to heredity and development. They offered explanations for the tragedy of birth anomalies and rationalized the occasional offspring that did not resemble the presumed father. They also provided a common-sense account of the inscription of social experiences and psychological processes on body and biology.

By no means merely the residue of folk belief, maternal impressions theories constituted a component of reigning scientific theories of generation, heredity, and embryological development up to the nineteenth century. In contrast to our understanding of heredity today as the transmission of information in the form of molecules, early embryologists conceptualized heredity as in part accomplished by the molding of traits during early development.[30] Seventeenth-century philosopher-naturalist René Descartes, for instance, believed that the fetus had its own heart and internal organs, but that the mother, through her impressions, formed its exterior parts. The "mother is the '*formatrix omnium membrorum exterioirum*,' [source of external form] regularly communicating images to the fetus through the umbilical arteries that serve to shape and imprint its visible body," he wrote.[31] Maternal impressions theories, then, were not merely explanations of the origin of birth anomalies. They also provided answers to questions of how offspring resemble parents, how sex is determined, how new features evolve in a lineage, and how fully developed offspring with the power of movement emerge from seemingly inert material.

Maternal impressions theories persisted into the nineteenth century, merging with theories of organic evolution that presumed the ability of organisms to change in response to environmental cues and to pass these changes on to their offspring. In his 1812 treatise on the evolution of species, French zoologist Jean-Baptiste Lamarck famously conceptualized bodies as supple reactive systems, able to harness metabolic and energetic structures to alter themselves through the habituation of faculties in adaptive response to the environment. Lamarck's picture of the body as highly plastic and his theory of the inheritance of acquired traits provided a broad and modern philosophical framework for understanding how parents' habits could be transmitted to their descendants, including how the environment of the mother's body might impress upon her growing fetus.[32]

In the nineteenth century, maternal impressions joined with theories of acquired and constitutional inheritance. Reigning theories of heredity and conceptions of the body and health asserted that parental constitution—mental and physical—is passed to offspring at conception and during gestation along with ancestral and species-typical traits. "Organic deficiencies and irregularities, no matter how produced in the parent—whether by accident or existing from birth—are liable to be revived in the offspring," read an 1857 obstetrical text typical for the time.[33] Mainstream medical doctrine held that parents bequeath to their children not only their ancestral germ plasm, but susceptibilities toward disease and tendencies in temperament through the state of their accumulated somatic condition at the time of conception and during early development.[34]

On the basis of these theories, physicians enjoined prospective mothers and fathers to prepare for parenthood through deliberate cultivation of intellectual acumen and character, as well as muscular, nervous, and digestive health. Cautioned the 1892 *Marriage and Disease: A Study of Heredity*, "Our bodily and mental development, as received from our ancestors and modified for better or worse by ourselves, is a certain heritage for our children. As we improve our condition mentally or bodily, so will our posterity be gainers, and as we degrade our natures, so shall our children suffer degradation."[35] Once established at conception, such constitutional attributes were thought difficult to counteract through environmental intervention. Without assiduous care, they could continue to pass from generation to generation.

United with Lamarckian theories of the inheritance of acquired traits, theories of the transmission of the stigmata of overall parental constitution presented a fearful picture of generations harmed by parental lifestyle. But they also created a formidable foundation for positive social prescriptions for marriage, procreation, and child rearing. The "man or woman with a bad family history may, by a steady and virtuous life, a strict observance of the laws of health, and proper care in the selection of a partner, live down the evil, so to speak, and leave an unencumbered estate to the children of the next generation," relayed *Marriage and Disease*.[36]

While nineteenth-century hereditarian philosophies recognized a role for both maternal and paternal impressions, special emphasis was laid on the imprint left by the pregnant mother during gestation.[37] *Aristotle's Masterpiece*, an eclectic pocket-sized work on the sciences of sex and reproduction, popular well into the nineteenth century, presented the case for maternal impressions with vigorous certitude.[38] Should a pregnant woman "fasten her eyes upon any object, and imprint it on her mind . . . it ofttimes so happens, that the child in some part or other of its body, has a representation thereof," averred the little book reprinted as many as one hundred times.[39] Hereditary resemblance, remarkable characteristics, and birth defects, the book claimed, could all be explained by maternal impressions.

Maternal impressions theories lived on in the nineteenth-century medical and scientific literature as well. Physician James Whitehead declared himself "not ashamed to confess my belief in the occasional originate of deformity and disease" through maternal impressions, citing several medical case reports, including an instance of hare-lip he attributed to a mother's observation of a corrective surgery for the same disorder during "utero-gestation."[40] Medical journals record a surprising number of published case reports of maternal impressions up to the turn of the twentieth century, evidencing the continued credibility given to the phenomenon. An 1888 *Journal of the American Medical Association* article, for instance, related the case of a "frog-like" or anencephalic child. While fishing in early pregnancy, the mother "caught a frog on the hook. It had swallowed the hook in such a way as to make it impossible to extract it. She attempted to kill the frog by crushing its head with a stick. As soon as she saw the mutilated creature she became deathly sick, and for days sickened at the thought of it." As the author concludes, "Though we may find no explanation, until we do we shall believe this monster due to maternal imagination."[41]

Giants of nineteenth-century German physiology earnestly propounded the theory of maternal impressions. The "celebrated embryologist" Karl Ernst von Baer, as Weismann referred to him in his discourse on the subject, even argued for the theory, credulously citing one case in which the visage of a fire affected "the mind of the lady so greatly" that "two or three months afterwards she was delivered of a daughter who had a red patch on the forehead in the form of a flame."[42] Karl Friedrich Burdach, author of "the received German text-book on physiology," had similarly argued that the mother's "'imagination influences the . . . homologous organs of the mother and the embryo" such that "when the former are disturbed a corresponding 'change in the formation of the latter may arise.'"[43] In varied forms, such theories persisted into the late nineteenth century and the eugenic age, renewed by calls to improve the human race through scientific management of human reproduction.[44]

Presenting the space between a pregnant woman and her growing fetus as an inscriptive, socially imbued environment with implications for the adult offspring and for future generations, maternal impressions theories were deeply entangled with reproductive and gender politics. On the one hand, they attributed great power and agency to the maternal body in reproduction and heredity and suggested an intimate and subversive female power that is resistant to masculine incursion, control, or knowledge.[45]

Some doctrines allowed that the mother could condition her own experiences and train her desires to enhance the outcome. Proper prenatal conduct through the training of emotions and the careful control of one's environment, according to this view, could not only prevent harm, but also improve the developing fetus. But this creative, agential reading of maternal impressions is complicated by the theory's prevailing conception of the maternal body as primarily passively conveying unwanted fates.[46]

Historically maternal impressions theories have overwhelmingly focused on the disruptive qualities of maternal impressions. Maternal impressions theories presumed and promoted a conception of the female body as porous, unstable, and irrational, a notion of femaleness long mobilized to argue for women's intellectual inferiority and lack of full claim to male rights, roles, and spaces. They relayed longstanding views that female bodies are uncontrollable, unpredictable, and unruly, and imputed that women are liable to be deceptive and worthy of suspicion, and to advance interests fundamentally at odds with the father's objectives. As such, as gender theorist Rosi Braidotti has argued, theories of maternal impressions were also "instrumental in creating, or strengthening, a nexus of stifling interdicts . . . on women."[47]

By the close of the nineteenth century, maternal impressions were increasingly disavowed in scientific writings.[48] Nonetheless, the theory remained sedimented in embryological doctrine and in the body of evidence underpinning the theory of inheritance of acquired characteristics. Writing in 1888, Weismann decried the continued belief in maternal impressions as folk theory defended in "form and language as scientific proofs."[49] Though resigned to the idea that "no one can be prevented from believing" in maternal impressions, Weismann insisted that "they have no right to be looked upon as scientific facts or even as scientific questions."[50]

In his sparring over the germ plasm theory and the inheritance of acquired traits, Weismann was compelled to take up the question of maternal impressions—for, as he observed, "there is a very close connexion between the theory of the efficacy of maternal impressions and that of the transmission of acquired characters, and sometimes they are even confounded together."[51] One such claim was a report of a pregnant sheep which, having broken its leg, birthed a lamb with a "ring of black wool . . . round the place at which the mother's leg had been broken, and upon the same leg."[52] Pointing to this example, Weismann argued that evidence offered in support of maternal impressions, just like claims of inheritance of acquired traits, takes the form of "post hoc" speculation, notoriously subject to "subsequent inventions and alterations" and the credulous patter of "old wives' fables."[53] A single coincidence of an idea of the mother with an abnormality in the child, he argued, should be rejected as spurious and unscientific evidence; hardly "proof of a causal connexion."[54] "Maturer knowledge of the physiology of the body," asserted Weismann, now obliges the rejection of maternal impressions theories.[55] Against the theory of maternal impressions, wrote Weismann, the "present state of biological science" holds instead that all traits are bestowed at the moment of fertilization: "with the fusion of egg and sperm-cell, potential heredity is determined."[56]

The Theory of Spermic Rejuvenescence

While the maternal contribution to heredity was seen as alarmingly pliable, risking the future of her child and even the human race, the male sperm was constructed as a heroic figure, charging up to the helpless egg to provide vigor and life-affirming strength to future descendants. Theories of the male sperm as the conveyer of rejuvenating or life-giving factors in heredity and development emerged from an old question about sex, newly recast in the mid-nineteenth century by microbiology and theories of evolution by natural selection: why are there two sexes at all?

The leading hypothesis was that sex—that is, the evolution of a second gamete, the sperm—evolved to "rejuvenate" a species. Writing in 1849, the British biologist Richard Owen hypothesized that "spermatic force" is required, in humans, "to set on foot the developmental processes," and in unicellular species that occasionally conjugate and exchange nuclear material, to refresh the lineage.[57] Referencing the Biblical story of Creation, Owen postulated sperm as the God-given source of fecundity and vital movement in the earth's creatures. "The Sacred volume," he wrote, "leaves us to infer that certain plastic and spermatic qualities of common matter were operative in the production of the first organized Beings of this planet." The female egg supplies "a mass of cells"; in contrast, the male sperm provides "the plastic force . . . in the formation and adjustment of the different tissues and organs of a new individual."[58]

Later rejuvenescence theories appealed not to Genesis but to Darwin's *Origin of Species*, and not to abstract "powers" and "forces," but to closely observed experimental evidence from the world of cytology and microbiology.[59] Belgian embryologist Edouard van Beneden is today remembered for his 1880s demonstration, in roundworms, that conjugation in meiosis involves a quantitatively equal pairing of the half-nuclei of the sperm and egg. Despite this, Van Beneden insisted that the male and female nuclear contributions are *qualitatively* unequal. Like Owen, Van Beneden believed that fertilization by the sperm functioned to rejuvenate the egg. Prior to beginning meiosis, he hypothesized that the egg is "hermaphroditic," meaning that it contains all the elements required to produce either a male or a female. During meiosis, the egg expels the "male principle." Later, upon fertilization, the sperm contributes a fresh male element, refurbishing the egg's ability to continue to divide.

Van Beneden's account seemed consistent with a broad base of newly uncovered evidence on reproduction in unicellular organisms such as bacteria. Single-celled organisms occasionally pair to exchange nuclear material prior to reproducing through cell division, a process known as conjugation. The leading explanation for the evolutionary purpose of unicellular conjugation during the 1870s and 1880s held that it provided a means of preventing senescence and revitalizing a hereditary lineage. Invoking a unity between the principles guiding conjugation in single-celled organisms and sexual reproduction in higher species, scientists suggested that two conjugating unicellular organisms performed like the egg and sperm at fertilization, and that sexual reproduction, too, functioned to "rejuvenate" the egg.[60]

Crucial evidence arrived with French microbiologist Emile Maupas' 1889 studies suggesting that microorganisms that failed to regularly conjugate degenerated and, over time, ultimately died out.[61] Maupas asserted that microbes that conjugated seemed rejuvenated and able to continue to robustly divide for many more generations.[62] Conceptualizing conjugation as a sexual union—as shorthand, he called the micronuclei involved in conjugation the "male pro-nuclei" and "female nucleus"—Maupas added a gendered valence to, as historian of science Frederick Churchill has quipped, "a process where there existed no other physiological or morphological indicator for distinguishing a male from a female 'gamete.'"[63] Of Maupas' findings, the authors of the 1889 volume *The Evolution of Sex* would fluoresce: "Conjugation is the necessary condition of [Protozoan] eternal youth and immortality. Even at this low level, only through the fire of love can the phoenix of the species renew its youth."[64] Maupas' theory of rejuvenescence through infusions of the male element suggested a thrilling continuity between conjugation in the world of unicellular organisms and fertilization of the egg by the sperm in sexual species: both, he claimed, served to maintain the vitality of the lineage.

Rejuvenescence theories find intellectual kinship in a deep well of conceptions of procreation as essentially monogenetic, involving the conveyance of the hereditary material of one parent, nourished by the other. Such theories derive from gender cosmologies that formed the foundation of premodern social life and that persist up to the present day in the language and symbolism of the gender system. Aristotle's fourth-century BCE account of sexual reproduction held that the male semen contained the vital element that confers movement, heat, and the potential for a sentient soul, while the female provided the nutritious material out of which the embryo is constructed. "'The male' and 'the female' are a principle,"[65] Aristotle wrote; the "male is the active partner, the one which originates the movement" while the female provides the "prime matter"[66]; "this in our view, is the specific characteristic of each of the sexes: that is what it means to be male or female."[67] In the symbolic cosmology of Aristotelian theories of male vitalization and monogenesis, males and females play different and asymmetrical roles in reproduction.

This belief system underpins long-running ideas that women merely contribute the "soil" in the service of "re-production," while strikingly resilient symbolism portrays men as transmitting the "seed," providing "the creative spark of life" and "essential identity of a child."[68] The doctrine reached its starkest form in seventeenth-century spermist preformationist doctrine, which held that the living organism exists fully formed in the male gamete. As Dutch microscopist Antoni van Leeuwenhoeck wrote in his first seventeenth-century report of observations of the sperm—which he called "animalcules"—under his lens: "I shall bring a sufficient proof of the fruits coming from the *Male seed*, and the *females* only contributing to the nourishment and growth of it."[69] The visage of the sperm moving under the lens of the microscope sustained conceptions that the male bestows animation, the spark of life, or individuality well into the nineteenth century.

When in 1828 the female egg was at last observed in mammals, shown to contain an identical nucleus to the sperm, and even demonstrated to emerge from the same gametogenic processes as sperm in early embryological development, the discovery did not, at first, entail a revision of monogenetic theories in which one sex, typically the male, conferred life.[70] Even as the concept of heredity itself was being transformed, nineteenth-century theories of the "rejuvenating" function of sexual reproduction and of the different roles of the egg and the sperm in transmitting hereditary characters continued to presume that the male and female provide asymmetrical and qualitatively different contributions to heredity, carrying forward the scientific and cultural presuppositions of monogenetic and vitalistic theories of the male as the carrier of the creative factor in heredity.

Weismann's commitment to sex equality in heredity stood in extreme dissidence to these deeply rooted ideas. "The same fundamental idea," Weismann wrote in his excoriation of rejuvenescence theories, "runs through all theories of fertilization up to the present time—the idea that the . . . 'vitalization of the egg' is the important part, or, as we may say, the true purpose of sexual reproduction."[71] He condemned Van Beneden and Maupas as "upholding a long-vanquished and mystical principle"[72] of reproduction as "a stimulus, as 'the spark in the powder cask,' or in biological language the vitalizing of the egg . . . directly derived from the old vital force of earlier times."[73] Such conjectures, Weismann charged, originated from longstanding and prescientific conceptions of the functions of maleness and femaleness within the reproductive order: "Thus according to Aristotle the father confers the impulse to movement, while the mother contributes the material," he recalled.[74] Contra Aristotle, Weismann argued that "there is no male or female principle, but only a paternal and maternal substance."[75]

Sex Complementarity in Heredity

Last was a third strain of nineteenth-century ideas about the asymmetrical role of males and females in heredity, situated within debates over the physiological basis of hereditary transmission. Like Weismann, many scientists turned to the physiology of the egg and the sperm in the race to locate the biological mechanism for the operation of evolution by natural selection. These aspirants to a biochemical theory of heredity, however, faced a puzzle: How do traits transit from generation to generation with high fidelity if they must pass through the highly differentiated male and female egg and sperm?

One prominent scientist who grappled with this question was the American zoologist William Keith Brooks. Today, Brooks is primarily remembered for building Johns Hopkins into a world-renowned center of biological research and for training the first generation of elite American biologists (including Thomas Hunt Morgan, who in his famous fly labs later unraveled the foundations of the gene theory). In his time, Brooks was also one of the leading American public intellectuals on matters of biology, society, and evolution. The disparate roles of the egg and sperm were central to his stance in the roiling 1880s debates over the physical basis of heredity.

Brooks' 1883 book, *The Law of Heredity*, was an "extended discussion and proof of the facts" in support of the view "that the ovum and the male cell perform different functions in heredity."[76] In the book, Brooks excoriated scientists—Weismann in particular—who "have either tacitly implied or directly accepted the view that the two sexual elements play similar parts in heredity," promulgating the erroneous proposition "that either parent may transmit to the offspring any characteristic whatever."[77] Studies of chromosomal behavior during early development of the sperm and egg had convincingly shown that male and female germ cells contribute quantitatively equal numbers of chromosomes at the moment of division. But, Brooks claimed, "at bottom, it is simply an assumption that the homology or morphological equivalence of the ovum and male cell proves their functional equivalence."[78]

Brooks believed, in line with Darwin's 1868 theory of pangenesis, that the cells of the body generate small particles called germs (or gemmules) to be passed to the next generation at fertilization.[79] In this process, Brooks argued, the egg and sperm play different roles. While the egg transmits the gemmules that form the embryo according to the species-type, the sperm, "as its especial and distinctive function," conveys the specific characters of the individual.[80] "According to this view," concluded Brooks, "the ovum is conservative; the male cell progressive. Heredity or adherence to type is brought about by the ovum; variation and adaptation through the male element."[81] Entering the fray of human gender relations, Brooks asserted that the political and social complementarity of male and female gender roles derives from the distinct functions of the egg and the sperm in heredity and evolution. Male and female "parts in the intellectual, moral, and social evolution of the race are, like their parts in the reproductive process, complemental," Brooks wrote.[82]

The intellectual qualities of men, he argued, mirror the male sperm's role in generating and retaining variability. "Being the variable organism, the originating element in the process of evolution, the male mind must have the power of extending experience over new fields, and, by comparison and generalization, of discovering new laws of nature."[83] Evidence for this, Brooks argued, could be found in the predominance of men as creators, abstract thinkers, and developers of civilization in human history. In contrast, females are "the conservative organism, to which is entrusted the keeping of all that has been gained during the past history of the race."[84] A woman's "mind is a storehouse filled with the instincts, habits, intuitions, and laws of conduct which has been gained by past experience."[85] Hence, women are suited to "occupations where ready tact and versatility" are of greater importance than education, and "where success does not involve competition with rivals."[86]

At pains to clarify that he did not think the sexes unequal, Brooks assured readers that "the sexes do not naturally stand in the relation of superior and inferior, nor in that of independent equals, but are the complemental parts of a compound whole."[87] Yet Brooks' account amounted to an affirmation of the natural justice of a status quo in which men dominated the realms of life bestowing personal, economic, and intellectual freedom. The data of biology, he concluded, affirms that "the positions which women already occupy in society and the duties which they perform are, in the main, what they should be if our view is correct; and any attempt to improve the condition of women by ignoring or obliterating the intellectual differences between them and men must result in disaster to the race."[88]

Brooks was not the only scientist to see a political threat in Weismann's assertion of sex equality in heredity. Scottish biologists Patrick Geddes and J. Arthur Thomson offered their 1889 book, *The Evolution of Sex*, as a "discussion and criticism of the speculative views of Professor Weismann," specifically, "the opinion maintained by Weismann" that "'the physiological values of sperm and egg-cell are equal.'"[89] Against Weismann, Geddes and Thomson argued that the sperm and the egg are fundamentally different "in spite of the structural resemblance in the rough features of nuclei."[90]

Maleness and femaleness, they asserted, express a deep and foundational duality in nature. In the egg and the sperm, "assuredly, the difference between male and female has its fundamental and most concentrated expression."[91] The life of a cell, they asserted, whether gamete or single-celled organism, involves dueling feminine and masculine processes: anabolism and katabolism [sic]. Anabolism refers to passive maintenance: "upbuilding, constructive, synthetic processes"; katabolism to activity and change: "a disruptive, descending series of chemical changes."[92] Though all cells contain elements of both, some, they asserted, have a greater anabolic "habit and temperament," while others are more katabolic.[93] Eggs are "passive, quiescent, enclosed, or encysted," like amoeba, while sperm are "active, motile, ciliated, or flagellate cells" with "predominant katabolism."[94] In unicellular microbes, Geddes and Thomson asserted, anabolic species are more species-maintaining, while katabolic species are more self-maintaining. So, too, in the sex cells: "Femaleness is anabolic preponderance in reproduction . . . similarly, katabolic preponderance stamps its character of active energy upon spermatozoon."[95]

Geddes and Thomson, like Brooks, did not hesitate to turn their biology to political prescription. Maleness and femaleness, and the sperm and egg, they believed, arise from a universal cosmological duality, a complementary opposition between katabolism and anabolism, foundational to all nature from simple Protozoa to human man and woman. Like the egg and the sperm, they wrote, the "two sexes are complementary and mutually dependent," reflecting a "deep difference in constitution" between males and females.[96] Katabolic males "are more active, energetic, eager, passionate, and variable"; anabolic females "more passive, conservative, sluggish, and stable."[97] Each human male trait was met by its female partner: males "lead," females "preserve"; males introduce "variation," females maintain "general" heredity; males exhibit "more intelligence," females "altruistic emotions"; males embody "independence and courage," females "affection" and "sympathy."[98] Like Brooks, Geddes and Thomson insisted that nothing about their perspective entailed the inferiority of women: "Each is higher in its own way, and the two are complementary."[99] Yet decrying "the complexly ruinous result" of the fight for women's right to vote and the entry of women into the workforce for "family life," Geddes and Thomson asserted that political debate over human gender roles must heed biology.[100] "The social order will clear itself, as it comes more in touch with biology," they contended.[101] To attempt to "obliterate" the differences between the sexes is to contravene "what was decided among the prehistoric Protozoa."[102]

"AMPHIMIXIS" AND THE THEORY OF THE GERM PLASM

Like his contemporaries, including Brooks, Maupas, and Geddes and Thomson, Weismann sought a material basis for the operation of evolution by natural selection in the biochemistry of the egg and sperm. But Weismann broke with their presumption that the egg and sperm must play different roles, analogous to perceived constitutional and complementary differences between the sexes. He dismissed complementarian notions of fertilization as "the union of two opposed forces"[103] as relics of "the unrecognized reflection"[104] of older and unsound ideas about the nature of sexual reproduction. Calling on scientists to "lay aside preconceived notions,"[105] Weismann pronounced the end of the "old doctrine of fertilization" presuming a "dynamic" between the male and female principles. "Fertilization," he argued, "has no significance except the union in the single offspring of the hereditary substance from two individuals."[106]

Abjuring references to "so-called 'sexual reproduction,'" Weismann

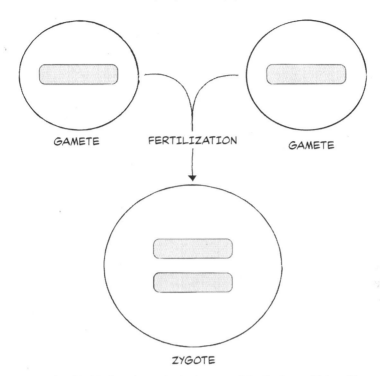

FIGURE 2.1 Amphimixis: Sexual reproduction is accomplished by the combining of the qualitatively and quantitatively equal genetic material of two gametes into a fertilized zygote. Figure by Amy Noseworthy, copyright Sarah Richardson, 2020.

coined a sex-neutral term for his theory of fertilization: "amphimixis" (amphi, meaning "both," and mixis, "mixing" or "mingling").[107] The theory that sexual reproduction is nothing more than amphimixis asserted that male and female germ cells contribute qualitatively and quantitatively equal hereditary material; with regard to heredity, they are functionally identical (fig. 2.1). The "essence of fertilization," Weismann wrote, is found "neither in the vitalization of the egg, nor in the union of two opposed polar forces," "but rather in the fusion of two hereditary tendencies,—in the mingling of the peculiarities of two individuals. The substances which come together in fertilization, from the male and from the female, are not fundamentally different but essentially similar."[108]

Hypothetically, according to amphimixis, if the nuclear substance of two sperm or two eggs could be combined, fertilization would have occurred. Indeed, Weismann's 1890 thought experiment suggesting that "if it were possible to introduce the female pronucleus of an egg into another egg of the same species . . . it is very probable that the two nuclei would conjugate just as if a fertilizing sperm-nucleus had penetrated" may stand as the earliest biological vision of the possibility of reproduction between two individuals of the female sex.[109]

As he developed his own radically opposing view of sex and heredity, the sharp-tongued Weismann relished pinning his opponent's views to outdated dogma, rhetorically aligning the germ plasm theory with modern ideas and against an older science laden with folk-theory-inflected concepts of the complementarity of male and female essences and misty cosmologies riven by deep miasmic dualities in the universe. Amphimixis—the doctrine of sex equality in heredity—Weismann implied, at last separates science from superstition. To the old, superstitious, and unscientific doctrine exemplified by Aristotelian doctrine and maternal impressions theories, Weismann contrasted his particulate, molecular understanding of inheritance grounded in the unsexed nuclear material of the reproductive cells.

Weismann was neither the first nor the only scientist of his time to assert the principle that male and female parents contribute equally to their offspring.[110] But it is Weismann who, by demonstrating the coherence of the equality of male and female hereditary material with the burgeoning cytological and embryological findings of his day as well as with larger theoretical developments in the sciences of heredity and evolution, launched the most vigorous and ultimately successful case for the essential validity of the doctrine that males and females contribute quantitatively and qualitatively equivalent material to the next generation.

Examining the evidence and reasoning that led Weismann to the germ plasm theory of heredity and the rejection of the theory of inheritance of acquired traits reveals that the hypothesis of sex equality in heredity was a construct developed in the service of cohering a broad range of findings from embryology and cell biology under the theory of evolution by natural selection. To use Weismann's metaphor, sex equality in heredity was a compass or, more provocatively, a miner's dynamite stick, not a bricklayer's stone. To say that the thesis of sex equality in heredity is a construct wholly entangled with and central to the germ plasm theory, of course, does not entail that it is "merely" a theory or construct, in the sense of being simply conjecture. Rather, it is to stress that the meaning of the doctrine—how each term, "sex," "equality," and "heredity," is to be understood—is embedded in the historically-specific context within which it was articulated.

SEX EQUALITY IN HEREDITY IN THE TWENTIETH CENTURY

A rising body of evidence in the early twentieth century soon heaped confirmation on Weismann's theory of the continuity of nuclear inheritance. A vivid example, cited as visual confirmation of Weismann's theory in biology textbooks until the mid-twentieth century, can be found in ovarian transfer experiments performed by William Castle at Harvard University.[111] Castle grafted the ovaries of a purebred jet black guinea pig into a purebred albino one, then mated her with an albino male. The albinos produced jet black offspring—insensate to the maternal milieu (fig. 2.2). Scientists interpreted this as definitive evidence for Weismann's claim that gametes retain their ability to determine specific traits with high integrity regardless of their intrauterine environment, even in mammals.

The galloping development of the study of the role of nuclear chromosomes in transmitting hereditary characters, leading to the chromosomal theory of heredity, gave impressive molecular confirmation and new precision to Weismann's hypothetical "germ plasm." Identical pairs of chromosomes from egg and sperm nuclei lined up in lockstep, divided, exchanged material, and transited sexlessly through the beat of generations, apparently without retaining markers of their maternal or paternal origins. Similarly, replications of Boveri's nuclear transplant experiments in many species proved, again and again, that the nuclear material alone was sufficient to produce a functioning gamete. Finally, the rediscovery of Mendelian laws of heredity in the first years of the twentieth century confirmed the equal capacity for transmission of most traits through either the maternal or paternal line. These developments gave a jolt to studies of heredity, snapping scientific workers into rapid consensus around a set of guiding principles that included the continuity of the nuclear genetic material and the essential equality of the maternal and paternal contribution.

However, Weismann's term, "amphimixis," intended to delink fertilization from sex, found few takers. Instead, the twentieth-century defenders and expositors of Weismann's germ plasm theory embraced

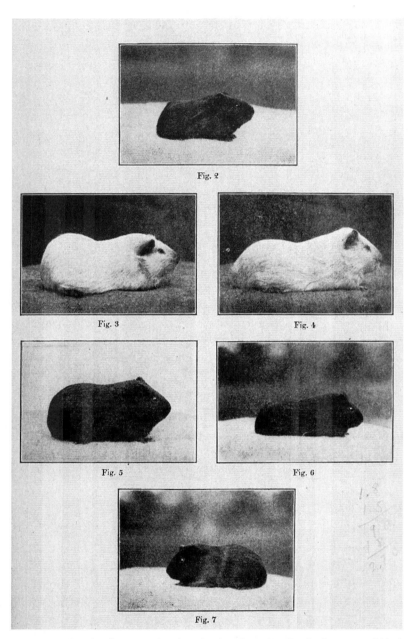

FIGURE 2.2 Results of ovarian transplantation in guinea pigs. Ovaries from a small black guinea-pig (fig. 2) were transplanted into an albino (fig. 3) that, mated with another albino (fig. 4), produced black young (figs. 5–7). From Castle, *Genetics and Eugenics*.

the more vernacular language of male and female "equality" of hereditary contribution. Introducing the nuclear theory of heredity in his 1896 *Cell in Development and Inheritance,* Wilson's opening gambit was Weismann's central premise: that the "paternal and maternal nuclear substances are equally distributed."[112]

The "equality" of maternal and paternal genetic contributions recurs as if a mantra in major expositions of the science of genetics in the first decades of the twentieth century. Princeton cell biologist Edwin Conklin's widely used textbook *Heredity and Environment in the Development of Man* stresses it repeatedly: "maternal and paternal chromosomes are . . . distributed with . . . precise equality"[113]; "always the division of the chromosomes is equal and non-differential"; cell division involves "a scrupulously equal" partitioning of male and female nuclear material.[114] It is similarly presented as cardinal doctrine in Harvard geneticist William Castle's textbook *Genetics and Eugenics*: "A majority of biologists, probably, at the present time believe with Weismann that heredity is due to material substances or determiners which are located in the chromosomes."[115] "The principal reasons for so thinking," Castle wrote, include "the very exact manner in which as a rule [the chromosomal material] . . . divides into two equal parts" and "evidence that the father is just as influential as the mother in determining the inheritance of children" despite the fact that "the egg-cell is vastly larger than the sperm-cell."[116]

Even decades later, in the revised and much-expanded 1925 final edition of Wilson's canonical text, the principle of sex equality in heredity retained its centrality in clinching the importance of the nucleus to heredity: "The nucleus undergoes an exactly equal and a meristic division," explains Wilson.[117] The lack of equality, in contrast, in the amount of cell cytoplasm contributed by the egg and sperm was now offered as evidence that it *cannot* carry the hereditary material. The division of the egg cytoplasm, wrote Wilson, "is often unequal, sometimes in an extreme degree . . . [with no] evidence of exactly ordered distribution of the division-products, such as is so strikingly displayed in the division of the nucleus." Such evidence, claimed Wilson, compels the now current "widespread acceptance of the nuclear theory of heredity."[118]

Theories of procreation are not merely scientific claims. As the anthropologist Carol Delaney has influentially argued, they enroll "such matters as how life comes into being; what it is composed of; who or what the agents are; what the person is, both male and female; and how persons are related to one another, the nonhuman world, and the cosmos."[119] To Weismann, we credit the insight that the hereditary material is contained to the nuclear germ plasm, and also, inextricably, the modern scientific consensus that the male and female hereditary contribution is equal. Against previous modes of thought, Weismann offered a refreshing and powerfully coherent new framework, enfeebling centuries of socially and ideologically valenced assertions of the inequality of the contribution of males and females to heredity. That males and females do not make different contributions to heredity—that, instead, heredity is wholly composed of the qualitatively and quantitatively equal joining of the nuclear material of the maternal and paternal reproductive cells—henceforward became a hallowed principle at the foundations of twentieth-century genetics and the sciences of sex.

In the realm of gender politics, in time it transformed the material terrain to which claims about the nature of human sex roles might appeal for scientific warrant. In the long view of the history of the sciences of sex differences, the victory of Weismann's theory of the continuity of the germ plasm accomplished the simply stunning destruction of parental impressions, monogenetic, vitalistic, and complementarian theories of male and female contributions to reproduction. Though today rarely appreciated, seen in this light sex equality in heredity is as great a break from all that had come before—from Aristotle to Geddes and Thomson—as its more-lauded companion, the theory of the continuity of the germ plasm. By the mid-twentieth century, sex equality in heredity was enshrined, alongside the theory of nuclear genetic heredity, as a premise of modernity and a badge of firmly materialist ways of thinking.

Uprooting remnants of monogenetic and patriarchal understandings of heredity and reproduction, the principle of sex equality in heredity shattered political ideologies of gender inequality grounded in women's and men's presumed asymmetrical contributions to heredity and helped clear the ground for new understandings of male and female gender roles. "Science by itself cannot give meaning," writes Delaney, in reference to the arrival of twentieth-century genetics showing that "both men and women contribute essentially and creatively to a child," "but it is a resource that can be drawn upon."[120] "Our notions of gender are so deeply involved with notions of procreation and biology . . . that changes in ideas about biology and procreation are bound to affect notions of gender. This is, of course, what has been happening. Some women, at least, have been learning that they are not merely vessels for the male seed, not merely nurturers and supporters of life, but co-creators and (perhaps more than) equal partners in this endeavor."[121]

GENDER, HEREDITY, AND MATERNAL EFFECTS

To grasp the inextricability of the doctrine of sex equality in heredity from the advent of the germ plasm theory of heredity is to appreciate the unease that resurgent theories of maternal effects in heredity would evoke for geneticists in the century to come. In genetics, it would become, like the famed "central dogma" of twentieth-century genetics that cohered scientific investigation around the expectation of a one gene–one protein correlation, a cardinal principle orienting investigation, but also a dictum setting boundaries on acceptable questions and repelling and relegating to the margins any anomalous scientific claims to the contrary.

Foremost among the victims of the dogma of sex equality in heredity were reprisals of any sort of claims that appeared to assert a distinctive or additional hereditary contribution by the mother—claims that became increasingly insistent as the science of genetics developed in the decades to come. By 1958, British reproductive biologist Anne McLaren, among the earliest to demonstrate maternal effects in mammals, would declare that although early geneticists had "seemed to offer the final and decisive proof of sex equality in heredity," studies of "maternal effects" now show that "the number of hereditary factors supplied by each parent" is not "of necessity equal," giving convincing evidence of "exceptions to the rule of sex equality."[122] Claims about maternal effects, McLaren well recognized, marked the return of a very old intellectual and social provocation, entering a charged debate about sex differences in contributions to reproduction and threatening the foundations of the germ plasm theory of heredity.

Despite its foundational status in genetics and celebration by some feminists, sex equality in heredity proves to have been, from the start, a more tenuous doctrine than generally appreciated. A historically sensitive reading of the meaning of the doctrine of sex equality in heredity at the turn of the twentieth century requires understanding it as a theory emerging out of longstanding debates over acquired characters and maternal impressions, vitalization and monogenesis, and the asymmetrical role of the male and female gametes in heredity and reproduction. It was also intimately embedded in the metaphysics and philosophy of science of the Weismannian reformulation of the basic theories of heredity and reproduction. Sex equality in heredity, that is, becomes a meaningful claim only if we accept Weismann's conceptual condensation of heredity to the moment of conjugation of the nuclear genetic material—and the theoretical suppositions that come along with this conceptual construct.

Though the continuity of the germ plasm and sex equality in heredity quickly became cardinal in mainstream genetics textbooks and among leading biologists by the early twentieth century, Weismann's ideas continued to be vigorously debated in a wider sphere. In reference to Weismann's picture of sex equality in heredity, three strains of critics draw special interest: partisans of the role of the maternal egg cytoplasm in inheritance, who objected to the nuclear supremacism of Weismann's germ plasm theory[123]; philosophical dissidents who took exception to what they perceived as the increasingly hard-nosed materialism of the sciences of heredity and reproduction, for which Weismann became an emblem; and progressive reformers deeply aligned with the vision of volitional human agency offered by Lamarckian evolutionary theory.

Indeed, Weismann's assertion of sex equality in heredity drew indignant protest from some. On the other side of the Atlantic Ocean, in the turn-of-the-twentieth-century Chautauqua tents and Methodist churches of the American Midwest, among early feminist intellectuals, in popular eugenics pamphlets and temperance movement literature, and across an intellectual gulf just as wide as the geographical one, updated doctrines of parental impressions claiming a wide berth for prenatal influences in heredity flourished under the banner of "prenatal culture." As we shall see, decrying the germ plasm theory and scorning the German scientist Weismann as a cold-hearted materialist in denial of accumulated human experience, prenatal culturalists argued that mothers, through their unconscious and conscious guidance of intrauterine development, are the supreme factor in heredity.

CHAPTER 3

Prenatal Culture

In 1911, the Chicago-based Correspondence School of Gospel and Scientific Eugenics took to the pages of *Physical Culture* magazine to promote its wares.[1] "Knowledge that will enable you to give your children a superb inheritance!" proffered its exclamation-laced advertisement. Showcasing endorsements from the president of the Women's Christian Temperance Union (WCTU) and the famed Clark University educational psychologist G. Stanley Hall, the school's teachings advanced eugenic sentiments typical of the time.[2] "Parents owe a debt to Society, the State, and to God, to produce the highest possible type of citizens," pronounced one of the school's seven principles of "intelligent parenthood and well born children."[3] But one novelty distinguished the school within the vernacular discourse of the eugenic era: its avid focus on prenatal influences. Customers were promised a diploma in "Intelligent Marriage and Parentage" through lessons in the "marvelous powers vested in Prospective Motherhood and Prenatal Influences."[4]

In contrast to the growing Weismannian consensus that hereditary endowment is confined to the nuclear germ plasm transmitted at conception, equal parts maternal and paternal, the school's doctrine of "prenatal culture" asserted a greater maternal part in heredity, via the womb. "Both in the law and the gospel of heredity, of the two parents, the mother has a far greater influence we believe firmly," wrote physician and prenatal culturist Emma Drake in her popular 1901 advice manual for women, *What a Young Wife Ought to Know*.[5] Specifically, prenatal culturists of the late nineteenth and early twentieth century claimed that influences transmitted by the mother during the intrauterine period are more important for the eventual character of the child than those transmitted by the father at conception. "Both in body and mind . . . her powers of transmission of habits, good or bad, are many times stronger than are those of the father," averred physician and eugenicist Mary R. Melendy, in her 1903 prenatal culture text *Perfect Womanhood for Maidens, Wives, Mothers*.[6]

The Correspondence School of Gospel and Scientific Eugenics, just one node within a field of resurgent ideas about maternal prenatal influences in heredity between 1880 and 1920, presents a concise microcosm of the intense transit between politics and vernacular science in progressive era America that supported the dissemination of prenatal culture theories. The school's founder, Mary Teats, was a "national evangelist"[7] for the WCTU from the late 1890s to the 1910s, lecturing on matters of family and sexuality to women-only audiences throughout the American West.[8] Her 1906 *The Way of God in Marriage* argued for the importance of woman's exercise of mental control during pregnancy "to give the best possible culture to that little life developing beneath her heart."[9] Chicago-based Blanche Eames, who served as assistant principal, brought a case for prenatal culture in her 1914 *Principles of Eugenics*. Women, Eames argued, should place "themselves under special training during . . . highly susceptible periods, in order to give certain desired tendencies of mind and character to their offspring."[10] The school board included Judge Ben Lindsay, of Denver, Colorado, a eugenicist and "authority on juvenile delinquency" who advocated for compassionate treatment of juveniles rather than punishment, as well as for birth control in marriage and legalization of mutual consent divorce.[11] Another Board member, Dr. Winfield Scott Hall, was president of the medical college of Northwestern University, and, from 1902 to 1910, headed the short-lived reformist American Academy of Medicine.[12] An advocate for sex education, Hall authored a physiology textbook and several general audience manuals on sex hygiene and reproduction, including *Dr. Hall's Sexual Knowledge* (1916) and *Sex Training in the Home* (1927).[13] No figure associated with the Correspondence School of Gospel and Scientific Eugenics, however, better represents prenatal culture in America between 1880 and 1920 than its chairman of the school board, the lecturer Newton N. Riddell, whose book, *Heredity and Prenatal Culture Considered in the Light of the New Psychology*, stands as the most elaborate statement of the theory.[14]

FOLK HEREDITY AND PRENATAL CULTURE

A remarkable forty-year trail of archival materials reconstructs the career of Newton Riddell, an American popularizer of the thesis that eugenic parenthood must begin before birth and especially during the intrauterine period. Riddell, who billed himself as a scientist, psychologist, and minister, traverses the period from the last bloom of phrenology, the nineteenth-century popular science of mind, to the eugenic age, animated by the twentieth-century science of genetics.[15] Tracing the sources of his ideas and their reception reveals the vibrant cultural life of updated theories of maternal impressions—the centuries-old belief that a pregnant woman's mental and physical state imprints on the growing fetus, producing hereditary variation—in progressive-era America.

Born in 1862 and raised on a Nebraska farm, Riddell took "an unusual interest in scientific and abstract subjects" as a boy, showing special fascination with the new science of electricity. His professional start came in the 1880s as an anti-alcohol lecturer in rural Methodist churches throughout Kansas and Nebraska.[16] In 1887, he earned a degree from the New York Phrenological Institute—a credential he would later reference as advanced training in "applied psychology."[17] Shortly thereafter he published his first book, *Human Nature Explained: A New Illustrated Treatise on Human Science for the People* (1895), a mix of phrenological doctrine and reflections on the sciences of mind, brain, and heredity that launched his lecture career.

Riddell defined prenatal culture as "the law whereby the acquired and the transient physical and mental characteristics of parents, particularly those that are most active for some time prior to the initial of a life, at the time of inception and in the mother during gestation are transmitted to offspring."[18] *Heredity and Prenatal Culture* promised to "enable parents to apply these laws to the improvement of their offspring" through "physical culture, brain building and soul growth . . . to endow each child with a good physical constitution, a well formed brain, a mind hungry for knowledge and a soul imbued with the principles of morality; so that in its birth they may give to the world a child of light."[19] Rejecting "Weismann's theory of 'the continuity of the germ plasm,'" Riddell argued that it is "extreme ground" to deny "the effect of acquired characters and maternal impressions."[20] In contrast to Weismannian doctrine, prenatal culture promised a role for "choice and volition" in heredity.[21] "By it," Riddell argued, "the otherwise inevitable results arising from the operation of the fixed laws of heredity—whereby like creates like—may be greatly modified. By its intelligent application, all unfavorable conditions . . . may be immeasurably improved."[22]

In his writings, Riddell advised parents to begin deliberate efforts at self-improvement in the interest of prenatal culture "at least a year before" conception and counseled the mother to "continue until the birth of the child."[23] Conception constituted the first critical moment in prenatal culture. According to Riddell, "initial impressions" from both parents' "physical and mental states" can "modify the heredity of off-spring" at the moment of conception.[24] Of highest importance, however, was prenatal culture during the intrauterine period. As such, prospective mothers require "even more preparation for parentage than the father."[25] During gestation, the fetus will be "continually subject to maternal im-pressions," as the "physical conditions and mental states of the mother during gestation—her impulses, emotions, joys, sorrows, thoughts and sentiments—make their impression upon the forming body, plastic brain and sensitive soul of her offspring."[26] Riddell counseled that through careful exercise and management of physical and mental states during pregnancy correlated to particular stages of embryonic development, the mother can prevent marks and deformities and shape the future traits and qualities of her child.

From the 1890s to the early 1930s, Riddell made his daily bread as a traveling lecturer and author. More than one hundred small-town news articles as well as archives of programs from the Chautauqua[27] circuits of the time suggest a fully booked speaker in high demand who gave hundreds or perhaps thousands of lectures—one 1908 article claimed 3,500 to date[28]—on heredity and prenatal culture, child rearing, psychol-ogy, early education, the brain, temperance, marriage, sex, and personal improvement and success. He performed in city halls, schools, YMCAs, Protestant churches, and tent Chautauqua festivals in small towns and penned at least ten books and numerous pamphlets. While one may presume his audiences were largely white, a positive review of his book appears in the Indianapolis-based African American newspaper the *Freeman*, praising it as "one of the best works dealing with the problems of heredity, procreation, Parental Culture, psychology, psychic phenomena and allied subjects," suggesting a wider multiracial audience.[29]

"Heredity and Prenatal Culture" was Riddell's signature attraction. One source described it as Riddell's "special lecture"[30]; another reported that "the subject is known as a hobby of Mr. Riddell."[31] Riddell claimed that its contents were based on deep scientific study and thousands of personally investigated cases.[32] The lecture remained a backbone of his oeuvre for thirty years. Riddell "considers that the factor of heredity and pre-natal environment is potent in solving many of the problems of civilization," reported a 1912 article in the *Rock Island Argus* headlined "Psychology of Heredity Theme; N. N. Riddell's Lecture on Prenatal Culture Best Yet Given; Cites Many Experiments." "He laid greatest stress upon the psychic factor as being the agent of transmission of acquired characters. 'Modify the psychology of the parent and you will modify the character of the succeeding life,' he said." In the course of the lecture, Riddell derided August Weismann's famous mouse-tail experiment, said to have disproved the theory of inheritance of acquired traits. Research by Elmer Gates—a prolific American inventor and self-proclaimed expert in "brain building," who carried out animal experiments on the effects of education on the brain and heredity in his Chevy Chase, Maryland, "Laboratory of Psychology and Mind Art"—had refuted Weismann's findings, according to Riddell. In "six generations," Riddell reported, Gates had produced mice "born with tails two and a half times as long as the normal mouse's tail."[33]

Touching on eugenics, temperance, the mind-brain-body connection, Christian sexual ethics, parenting and education, and self-improvement, Riddell's lectures on prenatal culture offered audiences a seamless narrative addressing a range of topics of the day with the promise of scientific insight. Reported the *Galveston Daily News*: "His lecture last night on 'Heredity and Prenatal Culture' . . . met with applause. Dr. Riddell stated that prohibition should obtain wherever the American flag waves. He also favored the teaching of eugenic laws, compelling men to study texts prescribed by the state before being granted marriage licenses."[34]

The aims of "the new science of eugenics," Riddell contended, could best be met through education in prenatal culture. Through prenatal culture—an idea "to which the public everywhere is becoming more and more alive,"[35] hereditary "determiners" can be "strengthened" and "corrected."[36] "Prenatal culture is no longer an experiment," averred Riddell in an Iola, Kansas, lecture in 1917, "but a fact demonstrable and of vital importance to society. It will not be long before all intelligent people will understand that education of a child should begin with its parents and be continued through the intra-maternal period."[37]

Turn-of-the-twentieth-century ideas about prenatal culture, not just in books, but also brought to life in the Correspondence School, in small-town newspapers, under Chautauqua tents, and in civic auditoriums, reveal a marked presence of maternal impressions theories—more so than generally appreciated—within major intellectual streams of progressive-era American thought. Vibrant social currents brought renewed interest in the special power of prenatal maternal impressions to shape offspring, updating the doctrine for the eugenic age: self-help philosophies promising individual uplift via the deliberate cultivation of the brain, body, and germ plasm; Victorian fascination with the power of the immaterial world, including proofs of the sway of mind over body; and burgeoning women's reform movements, including temperance, social purity, and suffrage, which sourced the creative power of the maternal-fetal relation as one basis for the need for women's sexual and political freedom.

The rebuttal of Weismann's germ plasm theory set a common cause among prenatal culturists and energized their work. They positioned prenatal culture as a formidable challenge to the germ plasm theory, charging that Weismann neglected the role of the environment and the power of the mind in altering or overcoming hereditary fate. In their defense of maternal impressions theories, prenatal culturists stood, more broadly, in opposition to what they perceived as the increasingly hard-nosed materialism of the post-Weismannian science of life.

EUGENICS AND PRENATAL CULTURE

Prenatal culture was, in the first instance, a vernacular scientific doctrine that offered a new version of maternal impressions theory suited to the eugenic age.[38] While today eugenics is frequently associated with hard hereditarian views of the sort that culminated in the mass sterilization policies and Nazi gas chambers of later decades, early twentieth-century eugenic thought embraced wide-ranging ideas about biomedical guidance of human advancement.[39] This included management of preconception health and the prenatal and postnatal infant environment. Many eugenics-era authors and experts emphasized the importance of social and environmental reforms. They asserted that eugenics entailed parental responsibility to produce ideal children not only through use of knowledge of the sciences of heredity, but also through the maintenance of their own physical health, including abstention from "poisons" such as alcohol, and through instruction of their children in values such as piety and civic virtue. Prenatal culturists' newly invigorated philosophy of hereditary improvement through prenatal conditioning represents one strain of these environmentalist and neo-Lamarckian approaches to eugenics.

Indeed, the eugenic period was far from one of accepted orthodoxies. Rather, it is better understood as a tumultuous time of contested ideas and unsettled medical authority within disruptive changes in the life sciences. Between the 1890s and the 1920s, the infrastructure of American science, medicine, and public health massively expanded. Universities and hospitals swelled with newly defined classes of professions, spawning broadened funding and shiny new laboratories. "Scientist" became a profession and life scientists in particular assumed ever-larger authority in intellectual and political life. Their knowledge was increasingly expected not only to heal but to convey metaphysical insight, mediate social disputes, spur economic development, and govern complex societies.

Prenatal culture discourse surged at just the moment that the genetic sciences of heredity were transforming the biomedical sciences. The powerful role of chromosomes, genes, and Mendelian ratios in heredity was being unraveled. Lamarckism, the view that parental traits acquired by experience could be transmitted to future generations, was under severe attack. The new genetics, conceptual offspring of Weismann's theory of heredity, asserted that the hereditary factors are sealed in the nuclear germ plasm, with no mechanism for transmitting traits acquired by experiences during an individual lifespan. These sciences were rapidly popularized into advice for everyday life from agricultural breeding to family planning. Medical specializations such as obstetrics expanded and pediatrics and prenatal care emerged during these years, significantly altering the landscape of authoritative scientific discourse about reproductive biology and transforming public perceptions of citizens' procreative responsibilities.[40]

Many Weismannian scientists and social reformers, flush with the conviction that human characters were determined by genetics, believed that the future of human civilization lay in the positive application of the principles of heredity for social improvement. Trends in the management of populations in an age of immigration, epidemics, urbanization, industrialization, and gender liberalization drew upon these new hereditarian ideas to landscape human potentialities and social futures. Eugenics-era prenatal culture philosophy represents a distinct sect within eugenic ideas of the period. "Prenatal culture" renewed maternal impressions theories with insights putatively drawn from new developments in the sciences of brain and heredity. It offered scientific methods of living that promised to produce offspring of ideal physical health and to cultivate the soul, morality, and Godliness in parents and future children. While American prenatal culture theories drew upon eclectic intellectual and cultural sources, two premises unite and distinguish this body of literature within the context of eugenics.

First, prenatal culture doctrine insisted on the supreme importance of the intrauterine period for race betterment. Dissenting from eugenicists who, influenced by Weismann's theory of the germ plasm, focused on improving the race by limiting reproduction among the unfit and encouraging it among those deemed desirable, prenatal culturists emphasized the power of maternal intrauterine influences for eugenic aims. In his 1897 *Maternal Impressions*, the Midwestern eugenicist and reformer Charles J. Bayer described prenatal culture as "heredity's twin and powerful sister"; in contrast to heredity, in which "Similar produces Similarity," "the mother's impressions—that is the state of her mind before the birth of her child" produces variation, "improvement," or "retrogression."[41]

Prenatal culturists shared the view that in addition to genetic inheritance, children begin life with a highly significant endowment provided by the prenatal environment, and that this environment is the product of life choices by the parents—the mother in particular—that can make the difference between a productive future adult and an undesirable one. "I believe that we may expect to have something like an art of rearing children—a science of eugenics! . . . The mothers of the civilized world will then be in possession of the data that will enable them to scientifically regulate the most sacred of all human functions by the light of biological and psychological science," pronounced Gates in his 1897 prenatal culture text "The Art of Rearing Children."[42] Wrote California suffragist and education reformer Georgianna Bruce Kirby in her 1882 *Transmission; or, Variation of Character through the Mother*, "Through the influence of happily elevating conditions surrounding and, as it were, pressing on the mother, *the children will be superior to the parents*."[43] "The destiny of the individual, hence of the nation, of the race itself, depends upon the mother and the prenatal conditions arranged for her child," wrote homeopath Swinburne Clymer in his 1902 *How to Create the Perfect Baby by Means of the Art or Science Known as Stirpiculture; or, Prenatal Culture and Influence, in the Development of a More Perfect Race*.[44]

Second, prenatal culturists denounced Weismannian theories of heredity and defended continued belief in the doctrine of inheritance of acquired traits. Prenatal culturists argued that traits molded during the intrauterine period represented a form of inheritance of acquired traits with implications not only for the immediate offspring but for the genetic lineage as well.[45] "All variation of character, physical and mental, takes place in fetal life," wrote Kirby.[46] As physical culturist and eugenicist Martin Luther Holbrook wrote in his 1899 *Homo-culture; Or, the Improvement of Offspring through Wiser Generation*, during "the period of gestation . . . the child acquires probably many of the characters which it inherits from its mother."[47] "This truth," he wrote, "obtains fresh significance when we consider that a woman's conduct affects the direction not only of her life, but the lives of her future children, and possibly of succeeding generations."[48] Drawing on theories of brain development and embryonic evolution, doctrines of educational psychology, and Christian- and New Thought–inflected metaphysical beliefs of the time, prenatal culturists insisted that through directed human effort, specific traits could be shaped prenatally from the moment of conception onward by the parents and passed on to future generations.

Adhering to the doctrine of inheritance of acquired characters, prenatal culturists believed that if applied generation after generation, mental modifications starting in the prenatal period could become hereditary.[49] Gates' experiments, oft-cited in the prenatal culture literature, were claimed to offer direct evidence for the plasticity and educability of the brain prior to birth and for the potential for hereditary transmission of maternally induced prenatal impressions. Training of particular faculties leads, Gates maintained, to "a far greater number of brain-cells," and this structural gain, it was held, could be passed on to subsequent generations.[50] Wrote physician, eugenicist, and prolific women's health author Melendy, "In illustration, the facts are cited that our race-horses are becoming faster and finer every year; that in England, stock-breeders study with the most painstaking care how to improve animals by judicious mating and breeding; that Japanese jugglers and acrobats are the finest in the world simply because they are trained to be acrobats for generations back."[51] In this same way, the "power" of prenatal culture "in the keeping of every mother," declared Melendy in her 1904 *Vivilore*, has the potential to create "a race of beings as far surpassing the present-day" (fig. 3.1).[52]

Most adherents to the doctrine of prenatal culture did not conduct or publish their own research or participate in the academic research enterprise. Neither a formal scientific theory nor an organized school of thought, prenatal culture appears as a commonly promulgated belief among American progressive-era authors offering practical advice to a general readership on heredity, sex, marriage, parenting, early education, and methods for self-improvement and success. But in this, prenatal culturists were not unusual. A vast literature of sociomedical texts hewed to similar form, mixing everyday intuition with speculative science in a manner designed for accessibility to a lay public even as they made attempts to be scientifically credible and systematic in their presentation, and cite and engage with reigning scientific theories, including, most prominently, those of Weismann.

That said, prenatal culture and maternal impressions theories were occasionally promoted in the early twentieth century even by more pedigreed scientific researchers and medical professionals. As we shall see, they also rose to the attention of leading eugenic scientists who excoriated the authors for perpetuating unscientific superstition and imperiling eugenic goals by directing women to the doctrine of prenatal culture rather than the selection of a eugenic mate. Thus, amateur though many prenatal culture authors may have been, their works constituted a formidable and influential lay science of prenatal influences in the eugenic

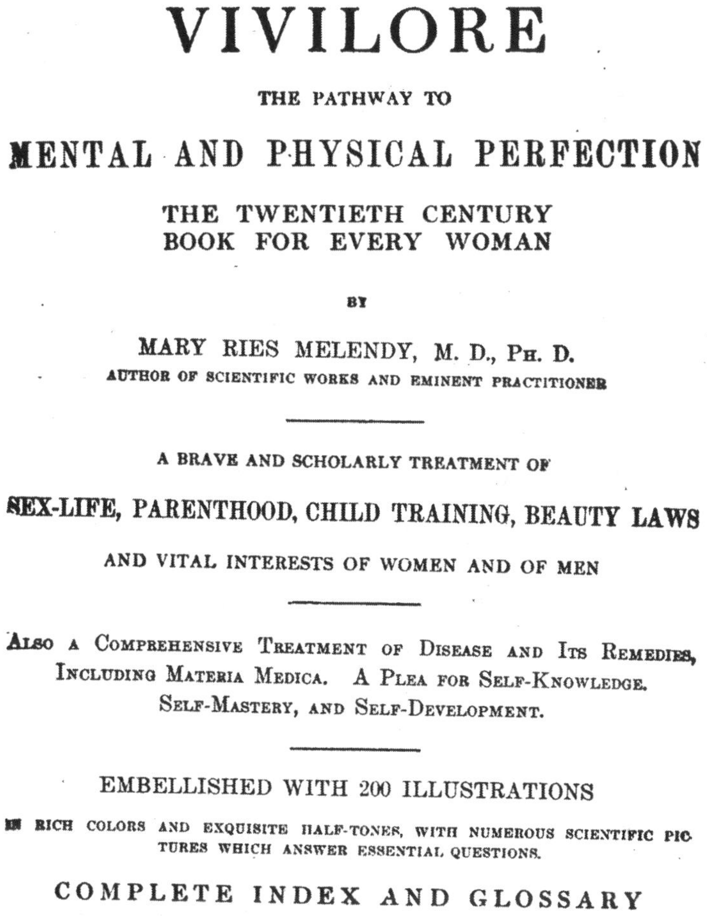

VIVILORE

THE PATHWAY TO

MENTAL AND PHYSICAL PERFECTION

THE TWENTIETH CENTURY
BOOK FOR EVERY WOMAN

BY

MARY RIES MELENDY, M. D., Ph. D.

AUTHOR OF SCIENTIFIC WORKS AND EMINENT PRACTITIONER

———

A BRAVE AND SCHOLARLY TREATMENT OF

SEX-LIFE, PARENTHOOD, CHILD TRAINING, BEAUTY LAWS

AND VITAL INTERESTS OF WOMEN AND OF MEN

———

Also a Comprehensive Treatment of Disease and Its Remedies, Including Materia Medica. A Plea for Self-Knowledge. Self-Mastery, and Self-Development.

———

EMBELLISHED WITH 200 ILLUSTRATIONS

IN RICH COLORS AND EXQUISITE HALF-TONES, WITH NUMEROUS SCIENTIFIC PIC-TURES WHICH ANSWER ESSENTIAL QUESTIONS.

COMPLETE INDEX AND GLOSSARY

FIGURE 3.1 Title page of Mary Melendy's *Vivilore*.

age. Prenatal culture offered a coherent vernacular scientific doctrine that found fertile ground in progressive-era trends, when the potential of each human life was thought to be limitless.

THE ORIGINS OF "PRENATAL CULTURE"

Appearing in American women's advice books, sex education materials, embryology and physiology textbooks, marriage and child rearing manuals, and works on heredity and eugenics, prenatal culture doctrine was not simply the unvarnished transfer of maternal impressions theories from the age of Paré and Descartes to the early twentieth century. It was a distinctive version of impressions theory tuned to the eugenic moment and built upon a set of transformations in the doctrine already under way in the nineteenth century.

Beginning in the eighteenth century, maternal impressions theories of the most crude sort were increasingly rejected by leading scientists. But this does not mean that they abruptly became the whispered folk convictions of the uneducated. Quite in contrast, the doctrine itself migrated and evolved to become a staple of nineteenth-century and early twentieth-century Anglo-American scientific theories of human heredity, parentage, mind, brain, and body, and the proper conduct of men and women. Parental impressions doctrine, infused with an optimistic social discourse promising moral and social improvement through cultivation of the mind and body, was, for many in the nineteenth century, an incontrovertible truth entirely of a piece with the theory that acquired characters could be transmitted to future generations.[53]

In the United States, the kinetic center for proto-eugenic ideas about the inheritance of acquired parental constitution and maternal impressions in the nineteenth century was phrenology and its allied engines of self-help, health, education, social reform, and well-being literature. Prior to the advent of eugenics in the late nineteenth century, phrenology was the most widely disseminated source of popular understanding of hereditarian and proto-eugenic ideas.[54] Through their storefront examination rooms, small-town lecture circuits, and accessible books and pamphlets, phrenologists presented a wide public with an appealing and intuitive theory of mental and physical health, integrating the theory of inheritance of acquired traits with a doctrine of moral and intellectual characters as products of specific regions of the brain, subject to growth, diminishment, and transmission to the next generation, similar to any other organic, physical trait.[55]

American phrenology most intimately followed a lineage of thought traceable to Scottish phrenologist George Combe. Contrary to common caricatures, phrenology did not depict the brain as hardwired. Combean philosophy presented phrenology as a system of "'cultivation' or 'training' of individual organs" of the mind in order to "favorably affect one's character."[56] In his 1829 *The Constitution of Man*, Combe promised the progressive improvement of humankind through knowledge of the physiological laws of mind and body.[57] Individual qualities of offspring, Combe argued, are largely determined by the temporary states of the parents just prior to and at the moment of conception.[58] Through deliberate shaping of one's own body and mind, one could improve one's hereditary lineage, overcoming ancestral blight and elevating the human race to new levels of perfection.[59]

Despite the invariable label of pseudoscience attached to phrenology today, the idea that there might be a science of education and self-improvement built on a proper understanding of human nature and the physiology of the brain became cardinal to the development of the psychological disciplines and arguably remains so today.[60] Liberal education reformers such as Horace Mann credited phrenological notions of scientifically informed guidance of child development as foundational to their educational philosophies.[61] Phrenology also influenced hereditarian thought. Major late nineteenth-century figures such as Herbert Spencer and Francis Galton acknowledged their debt to Combe.[62] In their proto-eugenic ideas, they cited central elements of phrenological hereditarian doctrine—in particular, the possibility of racial progress through proper mating and the optimization of one's reproductive health.

Phrenological ideas and institutions found their most enduring station in America. Phrenology's optimistic philosophy of individual and social improvement through scientific knowledge of the brain joined social reformist impulses and ideals of individualistic middle-class up-lift to form a powerful vernacular philosophy of health and self-help. In nineteenth-century America, phrenological philosophy offered reflections on human nature suited to the problems of the day and secular, brain-based advice for child education and adult self-improvement. Phrenologists argued for women's equality and suffrage, temperance and eugenics, and sex education and marriage reform, and urged the liberalization of child rearing and education.[63]

Over the course of the nineteenth century, an expansive genre of phrenological philosophical texts purporting to offer modern, scientifically informed advice to progressive men and women on sexuality, marriage, and parenting elaborated the role of maternal impressions in heredity and development.[64] In his 1870 *Creative and Sexual Science*, for instance, American phrenologist Orson Squire Fowler advised that for creating "superior and . . . inferior bodies and minds, ante-natal causes . . . are a thousand-fold more than all post-natal influences combined."[65] To the mother who asks "How can I carry my unborn in the very best manner, so as to write into its yet plastic nature all those intellectual capacities and moral excellences God has mercifully put within my power?," Fowler offered a severe and totalizing picture of the effect of even fleeting maternal states on offspring development.[66] "Each offspring takes on all those minor shadings and phases which appertain to its mother"; even "merely temporary states during pregnancy are also written right into the original qualities, mental and physical, of her offspring."[67] Mothers' "states of mind and feeling, while carrying their children, will be faithfully daguerrotyped, in all their shades and phases, upon those children, TO REMAIN THERE FOREVER, growing clearer and deeper as their existence progresses."[68]

The earliest English-language use of the phrase "prenatal culture" appears in the 1879 *Pre-natal Culture; Being Suggestions to Parents Relative to Systematic Methods of Moulding the Tendencies of Offspring before Birth*, published by the Washington, DC–based Moral Education Society.[69] "That a mother may, during the period of gestation, exercise some influence, by her own voluntary mental and physical action, either unwittingly or purposely . . . , in determining the traits and tendencies of her offspring is now a common belief among intelligent people," begins the first chapter, titled "The Law of Embryonic Moulding."[70]

A combination of parapsychological and phrenological doctrine underpinned *Pre-natal Culture*'s case for the physiological plausibility of maternal molding of offspring during the prenatal period. In his 1864 book *Man and His Relations*, the parapsychologist Samuel Brittan had posited that individuals can "electrotype," or copy the exact mold of, thoughts on their own and others' minds.[71] "The singular effect produced on the unborn child by the sudden mental emotions of the mother are remarkable examples of a kind of electrotyping on the sensitive surfaces of living forms," contended *Pre-natal Culture*, citing Brittan.[72] "If, then, we accept, as many do, the theory of modern phrenology, and regard the brain as made up of a congeries of organs, which are the instruments of distinct faculties of the mind or soul, it follows that if the mother during gestation maintains a special activity of any one organ, or group of organs, in her brain, she thereby causes a more full development of the corresponding organ or group in the brain of the fetus, and thus determines a tendency to special activity of the faculties, of which such organs are the instruments, in the child."[73]

Pre-natal Culture proposed a system of purposive direction of prenatal influences by the mother "to accomplish specifically-desired ends in determining the traits and qualities of her offspring."[74] In the absence of the mother's "voluntary and intelligent direction," it warned, the child will reflect the influence only of "chance-occurrences" and "parental or ancestral traits, good or bad."[75] Prenatal culture was hence a tool for overcoming the restrictions of parental heredity and reducing the prospect of fleeting exposures affecting an infant's development. Just as the body may be improved by directed exercise—"like the women of ancient Sparta exercised in gymnasiums in order to attain the highest bodily vigor, preparatory to the exercise of maternity," claimed *Pre-natal Culture*—so, too, might a "revived" practice today strengthen mental and moral muscle.[76]

Under the heading "Methods of Embryo Culture," *Pre-natal Culture* offered "a course of regimen and self-training" for parents.[77] To "confer vigorous muscles," the mother should engage in walking and "light gymnastics" "at the proper stage of pregnancy."[78] To develop mathematical ability in her future child, the mother should engage in exercises "in reckoning or calculating numbers"; by so doing she will "simultaneously increase the molecular deposits in that part of the fetal brain which is the organ of calculation."[79] Activity should be directed to counterbalance the hereditary weaknesses of the parents: "in any effort by a mother to cultivate her offspring in embryo through her own mental and physical action, she needs to give more especial attention to those desirable qualities, faculties or tendencies which may be deficient in herself, or in the father, and most especially such as may happen to be deficient in both."[80]

This system of prenatal culture embraced an intimate alignment between physical and moral health, emphasizing right living and mental and moral control as a method of organic self-improvement. The mother should avoid "all exercise of malevolent feelings, such as anger, envy, jealousy, hatred, revenge, covetousness, or wrong desire of any nature, since . . . such emotions, if indulged, may implant in the embryo the subtle germs."[81] She should also eschew "disagreeable and unprofitable associates" who might invite "untoward mental and moral influence . . . through the mother upon the forming child."[82] Through good conduct, parents can "cease to nourish the germs of physical disease and moral evil implanted in us by our progenitors."[83] Conversely, cultivation of physical health through diet and regimen can help eliminate "impure habits of thought and feeling."[84]

Although prenatal culture professed an optimistic organic philosophy offering "hope for all," implicitly, its advice presumed a readership with the money, freedom, and leisure to follow the instructions.[85] For those with the liberty to avoid disagreeable associates, hereditary weakness may be overcome, and through deliberate cultivation, hereditary traits may be altered and enhanced. "Improvement for the individual and for the race is possible," *Pre-natal Culture* promised. "In our ills and weaknesses, . . . inherited though they may have been from a long line of ancestry, we need not lie prone and helpless. . . . Help, purification, regeneration are within reach."[86]

CHILD CULTURE

The term "prenatal culture" likely took inspiration from "child culture," which first came into vogue in late nineteenth-century and early twentieth-century eugenic, kindergarten, scientific motherhood, and child welfare reform movements. A 1909 report of the National Child Labor Committee of the US Federal Children's Bureau attributes the term to the French: "Other countries, too, have awakened to realize the import of efficient guardianship of their children, have gathered expert information and are using it under the leadership of trained specialists. The French call this development 'child culture,' which implie[s] the use of scientific minds and trained powers, co-ordinated functions, and the protection of the state to the end of efficient manhood through a well guarded childhood."[87] Advocates of prenatal culture argued that such efforts should begin before birth.

"Child culture" signaled a modern, scientific, individualized, and child-centered approach to the social question of proper child rearing. Advocates believed that education should be rooted in the sciences of psychology, the brain, and evolution. In educational philosophy, child culture discourse was associated with a renewed assessment of children as in possession of rights, capable of moral learning, and deserving of the same privileges as adults. Writing of child culture in the *New York Evangelist* in 1876, the Reverend Frederick Clark decried the "chasm" between children and adults in the modern church and asserted that "childhood has its rights and duties" that should allow "children in the pews" to be "taught in the same words" as adults.[88] Perhaps the most powerful source of the spread of the term "child culture" in the late nineteenth century was the Kindergarten movement, popularized in English-speaking regions by the 1880 tome *Barnard's Child Culture*, which argued for progressive early education suited to the developmental stage of young children as a method of moral and mental development.[89] In late nineteenth-century scientific manuals for mothering, such as *Child Culture in the Home: A Book for Mothers* and *Child Culture; Or, the Science of Motherhood*, the term served to suggest a systematic and scientifically grounded approach to mothering in accordance with modern theories of early child development.[90]

In the broad psychological self-help literature of phrenology, authors adopted the term to refer to practical advice for early infant brain development and habit and character training. A "Child Culture" column appeared regularly in the *Phrenological Journal and Science of Health* during the 1890s, offering examples of well-cultured children whose parents had followed phrenological advice accompanied by detailed readings of the physiognomy of children (permitted by photographs submitted by parents). "The best mother," read the epitaph for the *Journal*'s Child Culture department, "is she who studies the peculiar character of each child and acts with well-instructed judgment upon the knowledge so obtained."[91] In the early twentieth century, socialist feminist Charlotte Perkins Gilman appealed to child culture theories in her classic excoriation of methods of child rearing in the nuclear family. Citing the need to "make provision for child-culture," Gilman decried the inefficiency, servitude, and, often, incompetence of exclusive and intensive child rearing in isolated domestic households compared to the prospect of its expert guidance in day cares and communal living arrangements.[92]

Alongside his expertise in heredity, Newton Riddell also claimed authority on the psychology of parenting and education, including in his own 1902 book, titled *Child Culture*.[93] He published articles in *Kindergarten Primary Magazine* and promoted his pamphlets with written endorsements from the prominent educational psychologist G. Stanley Hall.[94] At a 1907 Colorado Chautauqua program advertised to teachers, Riddell shared the stage with Hall. Riddell headlined the event with lectures "of special interest in Child Study and Education" and "heredity, prenatal culture, and various other phases of the child training problems."[95]

Appeals to the power of education and to ideas about brain plasticity in early child life were frequent reprisals of those who, at the turn of the twentieth century, resisted new hereditarian doctrines. These critics perceived Weismannians' denial of inheritance of acquired characters as a rejection of the facility of directed exercise, self-control, and education to alter an individual or a lineage's prospects. American popular science writer, horticulturist, and Neo-Lamarckian Luther Burbank, for instance, argued in his 1907 child culture tract *The Training of the Human Plant* that heredity can be modified and improved by healthy environment and gentle direction—and that such efforts will be amplified over generations. In this treatise on child education, based on extended analogy to cultivation of new plant varieties, Burbank argued that investing in the education of young, still-plastic children carried the highest dividends for the intergenerational improvement of hereditary endowments. While his focus in this text was on children, he suggested that the "mysterious prenatal period" may be "doubly" powerful for this aim.[96]

BRAIN-BUILDING

Prenatal culture writers represented their theories as an extension of these new educational philosophies. Reasoned Riddell, "If education is a factor in brain building and mental development, then education should begin when the brain is forming."[97]

For prenatal culturists, it was the brain, above all, that was to be the object of prenatal training. Wrote eugenicists T. W. Shannon and W. J. Truitt, "The mother has it largely in her power, *by the use of suitable means*, to confer on her child" a *"tendency of mind* and *conformation of brain."*[98] So-called new psychology—a term invoked by many prenatal culture authors—offered a philosophy of "brain-building" through directed exercises informed by the evolution, embryological development, and physiology of the brain.[99] To prenatal culturists and to educational psychologists such as Hall, the brain was like any other part of the body: its components could be strengthened through regular exercise. "Brain building is accomplished in precisely the same way that muscle building is, namely, by normal, systematic use," wrote Riddell.[100]

Like phrenologists of a previous generation, prenatal culturists held that specific skills, qualities of genius, and elements of character reside in the physical organ of the brain. Asserted Gates, "every conscious experience creates in some part of the brain a definite structure."[101] But prenatal culture authors distanced themselves from phrenology's parochialisms. "Phrenology had the misfortune of falsely locating every mental function," wrote Gates.[102] Now, maternal impressions theorists described the brain in terms of nerve cells, white and gray matter, electrical reactions, and developmental stages mirroring evolutionary phylogeny. Authors cited new evidence that the brain changes in response to training or injury and that it shows particular plasticity before birth.

Maternal impressions work, Bayer proposed, by affecting the brain and "the nervous system of . . . offspring before its birth."[103] Nerve cells and brain anatomy—the "plastic brain . . . of the forming child"—are the material mechanisms of maternal impressions.[104] Bayer hypothesized that "anything that makes an impression upon the mother's mind retards or promotes the normal growth of the brain cells."[105] As evidence, Bayer cited research on epileptics showing brain changes correlated with their symptoms: "Masses of compact fibrilla, or small fibres, were found in the gray matter of epileptics."[106] Scientific studies on how "nerve cells [are] contracted or enlarged," Bayer asserted, can be expected to similarly show "how the nerve cells are changed, destroyed, or arrested in their development" in response to maternal impressions.[107]

In ways not wholly dissimilar to today's hormonal theories of the action of maternal stress on a growing infant's future susceptibility to certain mental states, Gates, writing in 1896, argued that the mental state of the mother produces "poisons" in the blood that affect brain development: "irascible, malevolent, and depressing emotions generate in the system injurious compounds, some of which are extremely poisonous" while "agreeable, happy emotions generate chemical compounds of nutritious value, which stimulate the cells."[108]

Melendy, too, conceptualized prenatal culture as a form of brain training. Citing contemporary psychologists such as William James on mental training and the physiology of the brain, Melendy described thoughts, habits, impressions, and consciousness as embedded in the cells and structures of the brain and even visible to the eye through a microscope.[109] "Very recent discoveries in the operation of the human brain,"[110] Melendy related, show the specialization of the "nerve-cells of the brain" and that brains "vary in size" according to "the different faculties in one individual."[111]

Prenatal culturists believed the brain to be especially plastic during embryonic and fetal life. And, to a far greater extent than earlier doctrines of maternal impressions, prenatal culturists stressed women's control and agency, rather than passivity, in transmitting traits during gestation. "Prenatal impressions are more potential than postnatal. When the brain areas are forming it is possible for the mother by the assiduous exercise of mental powers to greatly modify the hereditary tendencies and to improve the mentality of her child," wrote Riddell.[112] "A prospective mother has the power to produce a brain and body such as she desires, limited only by her mentality and the limitations of nature," claimed Bayer in *Maternal Impressions*.[113] Through her efforts, the mother "educated upon the line of maternal impressions" can even "change or modify traits" given by the father.[114]

FEMINISM AND THE "CREATIVE SCIENCE" OF PRENATAL CULTURE

"Every pregnant woman should be considered as a laboratory in which she prepares a new being," wrote physician Annette Slocum, endorsing prenatal culture in her 1902 advice manual, *For Wife and Mother: A Young Mother's Tokology* (a term for the science of obstetrics).[115] Once characterized by early modern writers as an artisanal workshop, now the womb was a "laboratory," "wherein are fashioned the millions of denizens to inhabit the world."[116] During the prenatal period, the mother "must build canals, reservoirs, pumps, cylinders, flies, columns, domes, cellars, crucibles, retorts, ovens, and a multitude of chemical and mechanical laboratories of the most marvelous kinds; build telescopes, a telegraph and telephone system."[117]

To prenatal culturists, the mother was a scientist, an architect, and an artist, and the fetus her creation. Melendy's chapter on prenatal culture carries the title "The Mother-Artist (Pre-Natal Culture)"[118]; to "mould that new form and bring into being a perfect specimen of the human race," echoed Clymer, the pregnant woman "must be an artist."[119] "The body of the child in formation is [a] blank sheet of paper; the life led by the pregnant woman is the preparation of that paper, and lastly and greatest of all,' her thoughts and desires, held in heart and Soul, stamp that prepared body with what she wishes it to become."[120]

Asserting the supreme, creative, and agential role of women in shaping the next generation, prenatal culturists joined a rising wave of progressive movements pressing for greater social and political equality for women. Specifically, advocates linked the mother's powers in shaping her offspring's traits to calls for increasing sexual and domestic freedoms for women, access to education, and respect for women's intellect and autonomy of mind.'

Kirby's 1882 treatise on the role of the mother in heredity offers a ringing exposition of this agential, proto-feminist doctrine of maternal impressions. "The mother's office was, and is yet, by the majority held to be a secondary one and comparatively unimportant. 'She merely nourishes the germ given by the father' is the common supposition," wrote Kirby.[121] Even when the mother's role has been acknowledged, Kirby charged, the focus has been on the mother's ability to harm, not to improve. As Kirby wrote, it has long been "seen that the pregnant woman could affect the temper, the disposition of her child by yielding to angry emotions, but she was not credited with the ability to convey a serenity and sweetness of nature surpassing her own."[122] As society comes to realize a woman's "great maternal power" in the transmission of traits, Kirby predicted that "her character will broaden, her thoughts enlarge" and "subserviency . . . will no more belong to her than her brother."[123] By "self-control" and "exercise of her will," Kirby contended, a woman can prevent the inculcation of undesired traits in her offspring. Urging women to "actively" exercise abstract intellectual skills, Kirby suggested that women practice geometry and complex music each day during pregnancy—rather than engage in "household drudgery" such as "making jam" and "hemming skirts."[124]

The underlying argument was not entirely new.[125] A maternalist strain of proto-feminism had long deployed ideas about the social importance of women's motherly roles to argue for their political and intellectual freedom.[126] Pinning warrants for women's freedom and equality to maternity, though a sharply contested strategy among feminists through the ages, often proved highly effective in producing historic advances for women, including in the fight for American women's suffrage.

Mid-nineteenth-century phrenologist and early feminist Hester

Pendleton argued in her *The Parents' Guide; Or, Human Development through Pre-Natal Influences and Inherited Tendencies* that children inherit their qualities from both their mother and father, promising woe to "the gray-headed father," who, for lack of "cultivating . . . the mental and moral faculties of the wife of his youth," now finds himself ashamed by "the vices and follies of his children."[127] Too often, fathers claim that "education is of no use to women; I had much rather my daughters should know how to compound a good pudding than to solve a problem, or to cook a beef-steak properly, than to write an essay." This is shortsighted, for if women were better able to access education and intellectual pursuits to "devote themselves to 'self-culture,'" Pendleton asserted, they could imprint higher qualities of mind on their sons, and America would be better endowed with "men of genius."[128]

An increasingly progressive maternalism infused maternal impressions theories as the century advanced. Physician John Cowan took up Pendleton's banner in his 1869 exposition on parental impressions, *The Science of a New Life.* Invoking the power of prenatal influences, Cowan argued that women's "equality in freedom of thought and action, and rights in person and property equally with man" is essential to human evolution.[129] "These vital facts cannot be misunderstood . . . by any one who has carefully read and fully understood . . . the immense and almost unbounded influence of the mother on the destiny of the child during its pre-natal influence. . . . Endow a woman with the right of suffrage, the right to her own person, and the right to her own property . . . and, if she be a mother, the influence appertaining to the exercise of these rights will, in the life of the child, develop."[130]

But what should happen if parents seeking to prenatally produce a gifted poet, carpenter, or businessman in their growing fetus accidentally instill such qualities in a girl child? Cowan counseled parents to set aside worries of imparting girls with "a genius for an employment that is pre-eminently fitted for man." After all, she may need such employment. Certainly, she can excel at anything a man can do and "she will be admired for her genius and sought after for her ability."[131]

This vision of womanhood freed for maximally healthy and empowered motherhood—and of female children reared for self-sufficiency—reached its height in the turn-of-the-twentieth-century doctrine of prenatal culture. In his prescriptions for optimizing the power of maternal impressions through each stage of their pregnancy, Riddell devoted much of his guidance to the importance of women's freedom to exercise their minds and talents. To "transmit her special talents," women "must exercise them during gestation."[132] Riddell hence advocated women's "absolute freedom," "liberation," and "self respecting independence" as a crucial means to issuing well-born offspring: "If the mother is a slave, if she is compelled to subject her will to the will of the husband . . . rest assured that her child will be a slave, a born serf, lacking in self reliance, independence, sense of freedom, and the respect and dignity that belongs to the well-born."[133] As such, a pregnant woman "should . . . assiduously cultivate her own force of character. She should affirm, 'I am vigorous, I am free;' 'I have no fear of anything nor any one;' 'I will overcome every opposition.'"[134] Drake similarly urged women to "assert your individuality; study independence in thought and action; be self-reliant, self-contained" during pregnancy.[135] For optimal prenatal culture, Melendy likewise advised, "reading and study should receive most careful attention. Remember that the mother's superior culture in music, art, or any study, must be exercised during gestation, if it is to have any effect on the child."[136]

Maternalist ideologies, of course, carry tensions and contradictions. While Melendy called for more women to "lead hosts to battle for a nation's rights or for a noble cause, political, philanthropic or moral," she simultaneously warned women not to abandon their primary domestic duties, through which they might "help to raise the standard of purity and right living by making their home-lives so fragrant with beauty, peace and serenity that it is a benediction to enter their doors."[137] Despite their sentiments regarding the supreme power of maternal influences, not all prenatal culturists lined up behind growing movements for suffrage, higher education for women, and women's greater participation outside the home. "Intelligent mothers are essential. Intelligent, not in the sense of the present day higher education, but in the sense of knowing their duties and capabilities, when they become prospective mothers," wrote Bayer.[138] "Women rule through the men they have brought into the world, not by being career women; not by winning honors in social, political or national affairs," admonished Clymer.[139]

Nonetheless, many turn-of-the-century reformers invoked the theory of maternal impressions to raise the urgency of women's causes. As mothers whose condition could shape the gestating citizens of the future, these activists argued, women must be lifted out of poverty, relieved from backbreaking work, and supplied with modern health care and even means of birth control. The British birth control advocate Marie Stopes summoned the specter of a child damaged prenatally by her frightened or unprepared mother's desire for an abortion—an outcome that could be avoided with birth control.[140] The feminist temperance and sex activist Frances Willard, founder and president of the WCTU, leader of the National Prohibition Party, and women's suffrage advocate—who notably also set up house with a female domestic partner—raised the hereditary influences of women's prenatal behavior and exposures as a basis for prescriptions in favor of women's right to divorce.[141] Feminist prenatal culturists sought frank talk about sex. Citing the grave consequences of children gestated under maternal duress, they argued for the right of women within marriage to deny sex to her husband. They promoted freely engaged marriages built on love rather than an arrangement for sex, and fought to secure women's control over the timing and spacing of her child bearing. They even appealed to prenatal influences to argue for women's right to propose marriage—Riddell once made national headlines ("Grab a Husband!," read one) for making just such a scandalous suggestion.[142]

MORE THINGS IN HEAVEN AND EARTH

Prenatal culturists' visions of women's special role in bringing about eugenic race betterment through disciplined activity during gestation provoked outrage from many establishment scientific eugenicists. The revival of beliefs in maternal impressions, now trading under the term prenatal culture, led the editors of the *Journal of Heredity* to take up the matter for special excoriation in 1915. Beliefs that prenatal culture is "a short cut to race betterment" might "give a woman reason to think she might marry a man whose heredity was rotten, and yet, by pre-natal culture save her children from paying the inevitable penalty of this weak heritage," they wrote.[143] Women should know that they may further the goals of eugenics only by bearing offspring with strong genetic stock. The "future of the race" is at stake: "We have long shuddered over the future of the girl who marries a man to reform him; but think what it means to the future of the race if a superior girl, armed with correspondence school lessons in pre-natal culture"—an unmistakable reference to Teats and Riddell's school—"marries a man to reform his children!"[144] Concluded the editors: "No one, I venture to declare, has human progress more at heart than has the eugenist [sic]. It must, therefore, be to the interest of every eugenist to see that the superstition of maternal impression is driven out of existence."[145]

The California eugenicist Paul Popenoe similarly denounced prenatal culture in his 1918 textbook *Applied Eugenics*, asserting that "attempts to influence the inherent nature of the child, physically or mentally, through 'prenatal culture,' are doomed to disappoint. The child develops along the lines of the potentialities which existed in the two germ-cells that united to become its origin. The course of its development can not be changed in any specific way by any corresponding act or attitude of its mother, good hygiene alone need be her concern."[146]

Had the new Weismannian science of heredity shown maternal impressions theories to be mere superstition? The most explosive tensions surrounding prenatal culture doctrines—more so than their theories of brain development or elevation of women's agency in heredity and reproduction—emerge from this debate. Considered in this light, the flurry of prenatal culture texts at the turn of the twentieth century must also be appreciated as an outburst of resistance to the negation of wonder and human agency perceived in the antimetaphysical hereditarianism of Weismann's followers. To prenatal culturists, dismissive rejection of maternal impressions by academics represented evidence of Weismannian scientists' coldness, elitism, denial of common sense, and lack of empiricism. True scientists, prenatal culturists insisted, would test the plausibility of maternal impressions, rather than rejecting them as, on face, theoretically impossible—or as the unaccountable beliefs of mere women.

"The scientists of the age have been, and still are, busily engaged with their theories of 'ids' and 'idants,' 'gemmules,' 'physiological units,' 'biophors,' 'germ-plasm,' etc., vainly searching for a physical explanation for the phenomena of heredity," wrote Riddell.[147] Scientists who dismiss maternal impressions as "witchcraft and superstition," he wrote, are "dominated by . . . academic nonsense, empiricism and learned stupidity."[148] A regular "intelligent man or woman" could "contradict the professor and teach him a lesson that his books and theories had failed to teach."[149] It is the academic scientists, not prenatal culturists, who, in the service of their theories, had abandoned the search for truth. "If the materialistic theory of heredity will not admit of . . . maternal impressions, in the presence of thousands of well authenticated cases, an honest man should admit the facts and 'reform his creed.'"[150] In a 1913 Cedar Rapids, Iowa, lecture, Riddell likened Weismannian deniers of maternal impressions to a toad living at the bottom of a well: "What does a toad living in a well know about the world? He thinks it is perpendicular, tubular in form, water at the bottom, sky at the top, and moss around the edges. His concept is very different from that of another toad that hops along a sweet potato row on a June morning."[151]

Bayer's invective against scientists and medical doctors who denied maternal impressions was even harsher. Professional scientists, he charged, are "biased, full of preconceived ideas, and prejudices, from the fact that they have investigated man from one stand point only—the physical, through anatomical optics."[152] Like Riddell and other prenatal culturists, Bayer charged scientists with repudiating maternal impressions based on dogma rather than empirical investigation of the "indentations and undulations made by the stream of time, passing through the brain."[153] "Heredity is an overworked jade," used to "cover up ignorance." Scientists who rebuff maternal impressions, he railed, are a "class of parrot philosophers" suffering from "intellectual necromancy."[154]

In contrast to scientists "tenacious" in holding to "old theories,"[155] in his 1897 discourse on maternal impressions, Bayer presented a populist, puritanical empiricism: "In this investigation," Bayer begins, "no man's dictum, or dogmatic assertion, has been accepted because of his standing"; the case for maternal impressions proceeds on "common sense, reason and observation" alone.[156] Rebuking the Weismannian philosophy of science that privileged unifying theories grounded in physiological mechanisms over the brick-by-brick accumulation of observable facts, Bayer called for a science of maternal-fetal effects built from unadulterated lay experience: "It is urged that all preconceived ideas be laid aside; examine the facts, make your deductions from them, then interpret the facts which have been collected, and draw your conclusions."[157] Though some may deny the existence of maternal impressions based on "abstract reasoning," "dogmatic assertion," and "scholastic verbiage," "there is not an observing mother in the land but knows that these statements are in the main correct."[158]

Indeed, prenatal culturists defended not only the common man's right to knowledge based on his own scientific inductions, but specifically the claim of *women* to scientific knowledge. Eames decried those who would deny the knowledge of "prospective mothers" who "have made practical test" of maternal impressions and seen results. "When they declare, and others can perceive, that their efforts have been successful, it is only fair and just to admit the existence and effectiveness of the law of maternal impressions." "If, as Dr. David Starr Jordan says, 'Science is human experience tested and set in order,'" continued Eames, "then a woman who is a mother and is able to classify and analyze her experiences, is a true scientist; for it is facts, not theories, with which science deals; and a mother knows the facts of maternity as a man, even though he be a scientist, can not know them. It is for man to stand and wait, while woman has the experience."[159]

Within the medical establishment, too, the theory of maternal impressions became a flashpoint for turn-of-the-twentieth-century debates over the nature of scientific knowledge. "Whether maternal impressions do or do not influence foetal development, is a question still before the profession. Though we may not offer an explanation of maternal impressions, yet that is no argument against the existence of such influences," wrote one physician in the *Journal of the American Medical Association*.[160] Doctors defending maternal impressions theories appealed for openness to the scientifically unknowable, respect for the humble case report, and appreciation of the ineffable complexity of the maternal-fetal relation.

One doctor, writing in 1903, objected to those who would "relegate the maternal-impression theory to the medical lumber-room" as "a figment of the fancy." "There is a subtle and possibly a reciprocal connection between the mother and the fetus during the entire embryonic life." We must admit, he argued, that a pregnant woman's "make-up is intricate" and perhaps not subject to the normal rules of scientific inference: "She is a problem, to be solved by no simple rule of mathematics."[161]

An obstetrician writing in the *Lancet* in 1903 similarly addressed those who "scoff" and "deny the influence of maternal impressions," placing himself among those who "at the present day . . . are unable in the face of many well authenticated cases to refuse to believe that maternal impressions and foetal deformities do sometimes stand in the relation of 'cause and effect.'" Science, he suggested, may not ultimately have the answer. Perhaps, he concludes, maternal impressions must be considered real even though "the action cannot be explained pathologically": "There are more things in heaven and earth than are dreamt of in your philosophy," he concluded, invoking Shakespeare's famous admonition.[162]

Kansas physician Edwin Taylor Shelly responded to the *Lancet* authors with an impassioned attack on maternal impressions theories. Citing "the lamentable fact that popular twentieth-century obstetric textbooks written by highly respected twentieth century obstetric authorities still defend and uphold" the theory, Shelly excoriated maternal impressions as evidence of the continued influence of "the black cat of superstition" in medical practice.[163] He attributed the persistence of the belief to the "pall of empiricism," suggesting that in "the thraldom [*sic*] of an irrational empiricism," doctors draw their own conclusions from their limited therapeutic experiences, often leading them to perceive confirmation for beliefs such as maternal impressions.[164] Shelly rallied his colleagues to "demand so loudly and so urgently of the medical pulpit that it cease to preach this impossible dogma in order that the absurdity may soon be banished from our obstetric text-books."[165] In closing, Shelly decried the harsh weight of responsibility that such a theory throws on pregnant women: "Are not the burdens, apprehensions, and indispositions of the average sensitive, expectant mother already great enough without our making her wrongfully believe that she must also assume the awful, crushing responsibilities entailed by belief in the theory of impressionism?"[166]

As chapter 4 will show, critiques such as Shelly's ultimately prevailed. By 1920, theories of prenatal culture had exited the arena, to the furthest backwaters of alternative health advice, and guidance to parents on prenatal influences underwent a stark shift. Previously fixated on maternal influences during pregnancy, now advice centered on the preconception health of the sperm and egg. The result was a striking pivot in cultural and scientific conceptions of male and female prenatal reproductive risk and responsibility.

CHAPTER 4

Germ Plasm Hygiene

"What can parents do to determine before birth the character of the child? Nothing, so far as we know, except so to live that the germ cells are not enfeebled, and the foetus checked in its period of quiet nutrition." So wrote Stanford University geneticist and outspoken eugenicist David Starr Jordan in the *Journal of Heredity* in 1914.[1] According to this view, adopted with zeal and discipline by its adherents, heredity was contained in the germ plasm and the fetus protected by the placenta from all but the most severe violence.

This chapter journeys to a very different world than today's—a world in which males, as much as females, were presumed to shoulder prenatal reproductive risk. Old theories blaming women for all manner of malformations and scourges were rejected. In their place, a new and distinctive gender and reproductive biopolitics was ushered in, centered on what one physician-eugenicist called "the modern doctrine of the individual as the trustee of the germ-cells"—that is, the proper care and maximization of the potential of the germ plasm.[2]

The prenatal culturists had claimed that pregnant mothers can cultivate new traits in their offspring that would then pass to future generations. Scientific and medical eugenicists denied this. The agency of parents—both mothers and fathers—lay primarily in preventing the exposure of their germ cells to harmful toxins prior to conception. Turn-of-the-century eugenicists who ventured to offer prenatal advice to parents stressed the eugenic imperative of avoiding exposure of the germ plasm to "race poisons" such as alcohol, lead, tuberculosis, or syphilis. As University of Wisconsin geneticist Michael Guyer would counsel in the chapter on "prenatal influences" in his popular eugenics text *Being Well Born*: "Don't wreck a good germ plasm."[3]

This chapter explores the ideas of scientific eugenicists such as Michael Guyer, Auguste Forel, and Caleb Saleeby, who presented a spectrum of Weismannian views on the role of early environmental exposures in heredity. For scientific eugenicists who embraced a role for environmental factors in heredity and development, pre- and periconceptional exposures to the gamete and the germ plasm in the very early embryo, particularly on the part of the father, were of critical concern for the health and well-being of the next generation. By comparison, intrauterine exposures were of minor interest.

AGAINST SUPERSTITION: THE SCIENCES OF INTRAUTERINE INFLUENCES IN THE EARLY TWENTIETH CENTURY

In "The Truth About So-Called Maternal Impressions," physician Arch Dixon, writing in the journal of *Surgery, Gynecology and Obstetrics* in 1906, dismissed the theory of maternal impressions as "a relic of the Dark Ages and the offspring of superstition." Embryonic development, he insisted, unfolds of its own agency from the moment of conception, entirely removed from the influences of the mother. The placenta forms an impervious barrier between mother and fetus: "the foetal and maternal bloods *do not mix*, for none of the maternal blood escapes when the umbilical cord is cut, nor can the minutest injections through the foetal vessels be made to pass into the maternal vascular system, nor *vice versa*."[4] "Indeed, the setting hen, patiently hatching her chicks, is physiologically no more separated from her chicks than is the mother from her unborn child," wrote physician Edwin Taylor Shelly in 1907 in the *Journal of the American Medical Association*.[5]

Scientific advice to mothers followed suit. Addressing "the conviction [that] prevails among the laity that the character of a child depends greatly on the mother's surroundings during pregnancy," Johns Hopkins University professor of obstetrics Morris Slemons, in his 1912 *The Prospective Mother: A Handbook for Women during Pregnancy*, dismissed such theories as mere "superstitions." "Scientific investigation has brought to light," he wrote, that "departures from the usual form of the body occur during the earliest days of pregnancy and arise in consequence of some irregularity in the process which molds the body-form from a simple spherical mass of cells," well before maternal experiences can influence specific aspects of the child's development.[6]

In 1913, the newly formed United States Children's Bureau issued its first manual on *Prenatal Care*, a landmark in the development of the practice of prenatal medicine (which only became routine later, beginning in the 1930s). "Doctors and other scientists are now practically agreed," related the slim pamphlet meant for new mothers, that maternal impressions "have no basis in fact." The pamphlet related the scientific evidence: "there is no connection between the mother and the child in the uterus" and a "mother's blood never enters the child." "Nature in thus erecting a barrier between the mother and child," it concluded, "has specifically provided for the protection of the fetus from such injuries."[7] Similarly, the New York social worker and eugenicist Mary Read's 1916 *Mothercraft Manual* assured readers that "with the union of the two germ cells the inborn characteristics of the individual are determined, 'the gate of gifts is closed.'"[8] *Facts about Motherhood*, a 1922 pamphlet by S. Dana Hubbard, director of New York City Department of Health at the Bureau of Public Health, likewise advised: "Away with the bogey—there is no such thing as a mother's mark. Maternal impressions having physical harmful influence exist in the minds of uninformed ignorant individuals and such beliefs cannot be demonstrated scientifically."[9]

In assessing these statements, it is important to recognize that prior to the 1940s, the experimental and physiological sciences of mammalian maternal-fetal influences were sparse indeed. German descriptive embryology had, by the late nineteenth century, delivered a vivid and precise picture of the anatomy and physiology of mammalian embryonic and fetal development. This included an understanding of the placenta as an organ that provided a near complete buffer of the fetus from the maternal environment, which scientifically trained obstetricians such as Dixon and Slemons interpreted as definitive evidence against maternal impressions theories. Appreciation of the physiological role of hormones and of the placenta in mammalian growth and development, which would reveal a more profound interaction of mother and fetus than many denouncers of maternal impressions asserted, arrived only beginning in the 1930s. Furthermore, to experimentally sort out the plausibility of prenatal influences, scientists required the ability to manipulate mammalian reproductive physiology in ways still far beyond the capabilities of early twentieth-century scientists. This included, most critically, mastering the delicate and tedious art of egg transplantation, first achieved—haltingly—only in the 1940s and 1950s, as discussed in the next chapter.

The issues were not only technical. Before the mid-twentieth century, public health efforts in the United States focused overwhelmingly on the moment of birth and the immediate postnatal period. This made eminent sense, as high rates of maternal and postnatal infant mortality remained a public health emergency. As the obstetrician and historian of medicine Irving Loudon has documented, maternal and infant mortality actually *rose* between 1900 and 1930 in the United States, when nearly 800 women expired for every 100,000 births and up to one in ten infants died before the age of one.[10]

While the concept of prenatal care first began to be articulated in the early twentieth century, prenatal clinics came into wide use only in the 1930s along with the rise of hospital birth.[11] Even then, there were few effective and empirically based prenatal interventions or assays available— not even a chemical pregnancy test. The primary focus of prenatal care was medical supervision of the mother during pregnancy. Moreover, the sorts of large-scale resources and epidemiological methods required to empirically tackle the question of prenatal influences in humans would not emerge for several more decades.

Confidence in the impermeability of the placental barrier, technical limitations on the study of prenatal influences in human populations, and the justifiably greater focus on maternal and infant health during this period, however, only partly explains scientists' minimization of intrauterine effects. Simultaneously, the widespread acceptance of Weismann's germ plasm theory among geneticists and scientific eugenicists directed what interest there was in prenatal influences to the health of the egg and sperm.

SCIENTIFIC EUGENICS AND PRENATAL INFLUENCES

As genetics became accepted for studying the transmission of traits from parent to offspring, early twentieth-century scientists and physicians used their expanding social and epistemic authority to re-shape the bounds of acceptable public discourse about maternal impressions. "The new biology," as historian Hamilton Cravens has termed the burst of genetic hereditarianism at the turn of the twentieth century, was committed to a modern, experimentally grounded, and thoroughly material account of heredity.[12] Not only were maternal impressions theories wrong, but they were heretical and anti-science—they were superstition.

Asserting the primacy of the human germ plasm in determining all human traits, scientific eugenicists such as Princeton University biologist Edwin Grant Conklin advised women to reject "old wives' tales" about prenatal impressions. "Careful studies have shown that supposed 'maternal impressions' of the physical, mental, or emotional conditions of the mother upon the unborn child have no existence in fact," Conklin wrote in his 1915 *Heredity and Environment in the Development of Man*, a major early twentieth-century eugenics text that saw at least five editions, traversing instructional use in the classroom as well as general audiences. "From the moment of fertilization" the organism is "a distinct individual with particular individual characteristics" determined by the germ plasm. "Hereditary influences are transmitted only through the egg cell and the sperm cell and these influences are not affected by intra-uterine development," Conklin claimed.[13]

The eugenics movement was underpinned by a white ethnocentric vision of the future of Western civilization emphasizing racial purity, progress, and hierarchy. A "better race of men, with greater hereditary abilities, must be bred," Conklin asserted, typifying the frank racism and bald aims of human genetic engineering that characterized even moderate eugenic discourse. Without eugenic action, he warned that "barbarians or savages" who "have on the average more offspring than their civilized contemporaries" will fast outpace Western societies, consumed by women's "suicidal" demands for birth control and "freedom from marriage and reproduction."[14] Through scientific guidance, "the worst types of mankind may be prevented from propagating, and best types may be encouraged to increase and multiply." Only "sentimental regard for personal rights," stands in the way.[15]

As chapter 3 emphasized, eugenics was a diverse movement, and this was true even among Weismannian eugenic scientists. Paul Popenoe, editor of the *Journal of Heredity*, a founder of the California-based Human Betterment Foundation and a principal advocate for policies that would lead to the sterilization of an estimated 600,000 institutionalized Californians in the mid-twentieth century, believed that investment in prenatal care for parents was contrary to eugenic aims of improving genetic stock.[16] As he wrote in his 1918 *Applied Eugenics*, "There can be no doubt that sanitation and hygiene, prenatal care and intelligent treatment of mothers and babies, are truly . . . desirable. At the same time, as has been shown, these . . . measures result in the survival of inferior children, who directly or through their posterity will be a drag on the race."[17] Similarly, hardliners like Jordan suggested that parental alcohol use may be a eugenic salvation: by poisoning the germ plasm it reduced the number of alcoholics born in the next generation.[18]

But there were some eugenicists who believed prenatal influences to be an important realm for eugenic interventions. To build a picture of eugenic scientific thought about prenatal influences, I analyzed thirty book-length English-language eugenics texts identified as canonical by leading historians of eugenics. Founding texts, such as Galton's 1869 *Hereditary Genius*, made no mention of prenatal influences at all. Charles Davenport's 1911 *Heredity in Relation to Eugenics* referred to prenatal influences just once.[19] However, amongst the one-fifth of the texts that did address prenatal influences as a potential point of intervention for eugenic uplift, a remarkably coherent and crystalline message emerges. Within the canon of printed word generated during the pop-cultural heyday of British and North American scientific eugenics in the first decades of the twentieth century, writing on prenatal influences overwhelmingly concerned both parents' duties to the health of their germ plasm, rather than the mother's influences while pregnant.

With a research assistant, I reviewed approximately 60,000 lines of text from six books written by scientific eugenicists that addressed prenatal influences at length (table 4.1), coding them according to (1) the proportion of the text that discussed prenatal influences, broadly speaking, (2) whether those passages focused on intrauterine health or gamete health, and (3) whether advice addressed the behavior of male, female, or

TABLE 4.1 Scientific Eugenics Texts Featuring Prenatal Influences: Gametic/Intrauterine and Maternal/Paternal/Parental Effects as a Proportion of Total Text Devoted to Prenatal Influences

Scientific eugenic texts on prenatal influences	TOTAL TEXT DEVOTED TO PRENATAL INFLUENCES: Number of lines (% of total text)	PRENATAL INFLUENCES: INTRAUTERINE VS. GAMETIC		PRENATAL INFLUENCES: SEX OF PARENT		
		Intrauterine	Gametic	Maternal	Paternal	Both
Castle, W. E. *Genetics and Eugenics.* Cambridge, MA: Harvard University Press, 1916.	224.5 (2.0%)	6.7%	93.3%	9.6%	5.1%	85.3%
Conklin, E. G. *Heredity and Environment in the Development of Man.* Princeton: Princeton University Press, 1915.	517 (4.2%)	34.5%	65.5%	17.8%	1.7%	80.5%
Forel, A. *The Hygiene of Nerves and Mind in Health and Disease.* Trans. Aikins. New York: Putnam's Sons, 1907.	536.5 (5.3%)	15.5%	84.5%	5.3%	0.8%	93.9%
Guyer, M. F. *Being Well-Born: An Introduction to Eugenics.* Indianapolis: The Bobbs-Merrill Company, 1916.	615.5 (5.6%)	27.0%	73.0%	19.4%	5.0%	75.7%
Mott, F. W. *Nature and Nurture in Mental Development.* London: J. Murray, 1914.	405.5 (8.7%)	45.2%	54.8%	46.8%	17.7%	35.5%
Saleeby, C. W. *Parenthood and Race Culture: An Outline of Eugenics.* New York: Moffat, Yard, 1909.	844.5 (7.3%)	20.1%	79.9%	24.2%	11.6%	64.2%

both parents. We found that, first, scientific eugenicists concerned about prenatal influences mentioned gametic risks more often than intrauterine ones. Across the texts, 70 percent of passages on prenatal influences focused on gametes, while 24 percent treated prenatal intrauterine exposures (with the remainder unspecified). There was variation among the texts: in Castle's book, 93 percent of passages on prenatal influences were devoted to potential risks to the gametes, while Frederick Walker Mott's *Nature and Nurture in Mental Development* granted the least number of lines to gametic harms, at 55 percent. All authors, however, apportioned more space to gametic than to intrauterine health.

Second, scientific eugenic writing about prenatal influences did not single out mothers for instruction. Writers addressed both mothers and fathers. In Conklin's textbook *Heredity and Environment* 81 percent of passages on prenatal influences pertained to both parents, 2 percent fathers, and 18 percent mothers. In Saleeby's *Parenthood and Race Culture*, 64 percent of the material on prenatal influences addressed both parents, 12 percent fathers, and 24 percent mothers. In Forel's *Hygiene of Nerves and Mind*, a striking 94 percent of the discussion of prenatal influences was addressed to both parents, 1 percent exclusively addressed to fathers, and 5 percent to mothers. Mott's book stands alone in its emphasis on maternal factors, with 36 percent of passages offering prenatal advice addressed to both parents, 18 percent to fathers, and 47 percent to mothers.

In sum, when scientific eugenicists offered advice on prenatal influences, they focused on both parents' responsibility for the preconception health of the sperm and egg. It is there that eugenics-era geneticists saw the possibility for the direct influence of the environment on heredity and development—for good or for ill.

"RACE POISONS"

The term "race poisons" was coined by British eugenicist and obstetrician Caleb Saleeby, who, embracing ideas about white racial superiority and race purity that saturated eugenic discourse, suggested that toxic damage to the germ plasm could be as deleterious to a genetic lineage as Europeans interbreeding with "African Negros."[20] Through his prolific writings, including *Parenthood and Race Culture: An Outline of Eugenics* (1909)—described by one historian as the "first systematic book on eugenics," and which served, in Britain, as "the basic text for eugenics for some years"—Saleeby became a leading voice for eugenic public health reforms to limit environmental assaults to the germ plasm.[21]

Saleeby was a dedicated temperance activist, concerned with the effect of "racial poisons" such as venereal disease and alcohol consumption on human potential. He vigorously dismissed the possibility of so-called maternal impressions, asserting that there is "no nervous connection whatever" between mother and child. But he took issue with fellow eugenicists, such as Popenoe and Jordan, who rejected the need for attention to prenatal influences. Saleeby argued that all children should have the opportunity to reach their inborn potential. A focus on optimizing germ plasm hygiene through environmental interventions, Saleeby felt, was an approach to advancing eugenic aims superior to harsh policies to restrict reproduction.

Other "race poison" eugenic activists include scientists August Forel and Michael Guyer. Forel was a psychiatrist at University of Zürich, temperance activist, and leading figure in the international eugenics movement. Considered the father of Swiss psychiatry, in 1890 Forel founded the Abstinence Society, which later became the International Society for Combating Indulgence in Alcohol.[22] He lectured across Europe on eugenics and heredity, becoming well known for his memoir describing his fight against alcoholism. A zealous eugenicist convinced of the urgent need to protect the germ plasm from degenerative race poisons, Forel even supported the use of castration to prevent alcoholic men from reproducing.[23]

"As parts of our bodies the ova and spermatozoa can not take care of their own hygiene," Forel asserted.[24] Lifestyle habits and disease, he argued, could lead to "degeneration" of the germ plasm. "Every influence through which the germ is poisoned or otherwise injured, and which thus lays the foundation for inherited weakness in a healthy stock," he wrote, "can be called germ-corruption."[25] Responsibility lies equally with the mother and father. "Nutritive relations in the body of the mother are undoubtedly important for the normal health and development of the embryo; but they do not determine its individual characteristics in the slightest," wrote Forel, "because in spite of the very important influence of the mother, it receives, on the average, as many characteristics from the tiny paternal cell as from the ovum."[26]

Forel's driving concern was the dysgenic effects of alcohol consumption. He posited that alcohol use, by harming the egg and sperm, caused "social poisoning" and degeneration. "Acute and chronic poisoning from alcohol in particular, assuredly affects the reproductive glands and makes the germs there degenerate, so that the next generation is more or less crippled according to the extent of the social poisoning," he wrote.[27]

In North America, Michael Guyer, geneticist and chair of the Department of Zoology at the University of Wisconsin who was best known for his 1916 text *Being Well Born*, also took up the cause of race poisons.[28] *Being Well Born* presented a primer on the science of genetics. The text reached a broad popular audience and was widely used in the United States as a college-level textbook in the life sciences.[29]

Guyer professed a moderate eugenics that emphasized social reform and the importance of both nature and nurture to optimal human development.[30] Germ plasm hygiene lay at the heart of Guyer's discussion of prenatal race poisons. Guyer instructed parents in their "one sacred obligation to the immortal germ-plasm." "It is the sacred duty of every individual," he wrote, "to see that the maximal possibilities of his own germ-plasm are not lowered by vicious or unwholesome living."[31] As "trustees" of their eggs and sperm, parents are responsible to "hand it on with its maximal possibilities undimmed by innutrition, poisons or vice."[32]

HEREDITY AND THE DIRECT INFLUENCE OF THE ENVIRONMENT

In their emphasis on preconceptional environmental risks to the germ plasm, Saleeby, Forel, and Guyer represent a small but vocal segment of eugenics-era scientists, physicians, and public health reformers broadly concerned about the possibility of corruption of the germ plasm by environmental toxins. In 1909, eugenicist Roswell Johnson of the Cold Spring Harbor Laboratories, a central institution in the American scientific eugenics movement, called for eugenic "recognition of the capacity of environment to alter the germ-plasm" in an article titled "The Direct Action of the Environment." Lamarckian ideas had once compelled "social legislation aimed at the improvement of the environment and the prevention of the individual's abuse of himself," noted Johnson. Though Lamarck's theories are now rejected, the foundation for such public health agendas, Johnson argued, is "amply replaced by the newer doctrine of the direct modifiability of the germ-plasm."[33] Environmental exposures ranging from alcohol abuse to "the opium or cocaine habit, ill-ventilation, over-fatigue, under-feeding, or ill-balanced diets," Johnson argued, may "unfavorably affect the germ-plasm" and require address through "restrictive and ameliorative" social reforms.[34] Johnson's reflections led one 1911 *Journal of Heredity* author to declare that "direct action of the environment on the gametic cells" is "from the standpoint of national eugenics . . . of the greatest importance to study."[35]

Brave New World, the 1932 Aldous Huxley classic today routinely assigned in high school classrooms as an accessible cautionary tale about the future of genetic engineering, paints a terrifying picture of a dystopian caste-based society governed by eugenic reproductive engineers. But it bears recall that the method of producing classes of people to serve particular social functions that Huxley imagined was actually not genetic engineering as it is understood today. Instead of selection or repression of desirable genetic traits through manipulation of the nuclear genetic material, Huxley's eugenic technocrats undertook carefully timed gametic and early embryonic environmental exposures.

Cloning embryos is "mere slavish imitation of nature," declares the fictional Mr. Foster, Director of "Hatcheries and Conditionings," in the opening passage of *Brave New World*. The "human invention" comes in the prenatal manipulation of the environment of egg, sperm, and embryos to "predestine and condition" them "as future sewage workers or . . . future Directors of Hatcheries." Mr. Foster proudly exhibits how the hatchery reduces the amount of oxygen to the brain for the lowest-of-the-low, the Epsilons. Other embryos, destined to be miners in the tropics, are given "heat conditioning" so that they will "thrive on heat." At the embryonic stage at which they "still have gills," they receive inoculations for tropical diseases such as sleeping sickness. Future chemical workers are dosed with "lead, caustic soda, tar, chlorine." Those slotted to be rocket-plane engineers are incubated in rotating containers "to improve their sense of balance."[36] Fiction though it was, Huxley's futuristic vision of producing calculated strengths and weaknesses in a scientifically engineered citizenry via early chemical and physical exposures may have been inspired by some of the experiments of the day.

For scientific eugenicists, the question of prenatal influences concerned not maternal effects during pregnancy, but toxic exposures to the germ plasm itself. A substantial stream of embryological and genetic studies investigated the consequences of "direct action of the environment" on the sperm, egg, and earliest stages of embryonic development. (Note that germ plasm is referred to in this literature in two senses. There is first the germ plasm as it resides singly in the adult parents, and then there is the germ plasm as it resides, paired, in the developing offspring. Both were to be protected and nourished.) In *Heredity and Environment*, Conklin summarized the available scientific evidence on the germ-enfeebling effects of an impressive array of harsh exposures:

1. Physical stimuli including the following, (a) mechanical, (b) thermal, (c) electrical, (d) radiant, (e) light, (f) density of medium, (g) gravity and centrifugal force, etc.
2. Chemical stimuli include the action of (a) substances found in normal development, such as oxygen, carbonic acid, water, food, secretions of ductless glands etc., and (b) substances not found in normal development, such as various salts, acids, alkalis, alcohol, ether, tobacco, etc.[37]

In the wake of Weismann's germ plasm theory, eggs and sperm from frogs, insects, and guinea pigs were subject to immersion in heat, cold, acid, alcohol, and all manner of imaginable environmental exposures to discern effects on fertilization and development, and to assess the possibility that environmental lesions to germ plasm could be passed from generation to generation.[38]

Although many well-known eugenic-era popularizers of findings of direct environmental modifications of the germ cells, such as Jacques Loeb and Paul Kammerer, were anti-Weismann neo-Lamarckians, such experiments were not necessarily heterodox to the germ plasm theory.[39] As Weismann wrote in his 1885 essay proposing the theory of the continuity of the germ plasm: "I am . . . far from asserting that the germ-plasm . . . is absolutely unchangeable or totally uninfluenced by forces residing in the organism within which it is transformed into germ-cells. . . . It is conceivable that organisms may exert a modifying influence upon their germ-cells, and even that such a process is to a certain extent inevitable."[40]

In fact, mutations caused by direct environmental effects on the germ plasm formed a key element of Weismann's germ plasm theory of natural selection and evolutionary adaptation to the environment. "If we trace all the permanent hereditary variations from generation to generation back to the quantitative variations of the germ . . . they can be referred to the various external influences to which the germ is exposed before the commencement of embryonic development," he wrote.[41]

Weismann also offered the "direct influence of the environment" on the germ cells as the singular means for the inheritance of so-called acquired traits. Plausibly, he conjectured, exposure to heat, disease, or chemical stimuli could alter the germ cells. This would account for some cases of apparent inheritance of acquired traits. In his mid-1880s essays, Weismann cited evidence that diseases such as syphilis and tuberculosis can be transmitted through the germ cell itself.[42]

To be sure, Weismann dismissed most claims of direct environmental effects as "doubtful and insufficiently investigated"[43] and in any case "very slight."[44] Direct environmental effects, he argued, were most likely to lead to general impairment or even nonviability, rather than the acquisition of specific traits. One might acknowledge that "external influences" could lead germ cells to be "arrested in their growth," wrote Weismann, but "this is indeed very different from believing that the changes of the organism which result from external stimuli can be transmitted to the germ-cells and will re-develop in the next generation at the same time as that at which they arose in the parent, and in the same part of the organism."[45]

As historian of evolutionary biology Peter Bowler affirms, early twentieth-century experiments showing that "deleterious external influences merely damaged the germ plasm in some way" were "a long way from the original Lamarckian claim that positive new characters could be built up in the germ plasm in response to adaptations acquired by the adult body."[46] Conklin in 1915, for example, wrote that while "there is no doubt that the germ plasm although very stable can and does change its constitution under some rare conditions," this presents no fundamental challenge to the germ plasm theory, which is "in the main . . . accepted by the great majority of biologists."[47]

Nonetheless, the possibility of direct environmental influences on the germ plasm allowed "race poison" theorists to profess themselves orthodox Weismannians while advancing the possibility of intergenerational harms caused by the exposure of the germ plasm to toxins. They took diverse approaches to this.

Coining the term "blastophthoria," which he defined as "germ corruption," Forel argued that prenatal environmental influences could threaten a genetic lineage.[48] Blastophthoria refused a hard distinction between "inherited" and "environmental" factors. "Diseases of the germ or *blastophthoria*," Forel wrote, exist "midway between inheritance and adaptation."[49] Borrowing from the eccentric German theorist of memory and heredity Richard Semon, Forel advised that to prevent blastophthoria parents should take up exercises to strengthen and train the nervous system, which may, he stressed, *over long periods of time,* "affect the germ cells," making possible "an exceedingly slow inheritance of acquired characteristics after innumerable repetitions . . . without impeaching the correctness of the facts adduced by Weismann."[50] After generations, "a very healthy and normal mode of life gradually improves the quality of such a bad breed."[51]

While Forel thought of prenatal environmental influences as an intermediate form of inheritance, Saleeby held a harder line, asserting that "each generation makes a fresh start."[52] For Saleeby, environmental factors were strictly complementary to hereditary ones, not a component of them. The "acquired deterioration of the parents . . . is not transmitted to their children," he wrote.[53] With social and environmental reforms, Saleeby argued, every good germ plasm might avoid harms and reach its full potential. Yet Saleeby also suggested that "racial poisons" could compromise innate potential, with intergenerational consequences: "Acquired characters are not transmitted; but the racial poison makes dim the lamp ere he passes it on."[54] Racial poisons, that is, could lead to reduced vigor or overall health in a lineage.

Guyer, too, professed allegiance to the germ plasm theory of inheritance, making clear that the consequences of poor environment could have hereditary implications only if they affect the germ plasm.[55] But he seems to have had a far-reaching conception of the harms of race poisons for a genetic lineage. "Impaired vitality" caused by poor prenatal environment, he wrote in *Being Well-Born*, "factors in the birthright of the child, who, moreover, may bear in its veins slumbering poisons from some progenitor who has handed on blood taints."[56] With colleague

Elizabeth Smith, Guyer even attempted to demonstrate the influence of germ plasm "blood taints" across generations with a series of experiments in rabbits. He exposed rabbits during early development to blood from chickens immunologically sensitized to rabbit eye tissue. He then bred the damaged male offspring with unaffected females. The results, he claimed, demonstrated that the resulting eye condition could be passed on through several generations, a case of "true inheritance"—not "merely placental transmission . . . from the blood stream of the mother"—passed "through the germ-cells of the male."[57]

PRENATAL ADVICE TO MEN

Viewed from our present moment, a striking feature of scientific eugenicists' approach to the question of prenatal influences is its egalitarian view of the importance of both male and female prenatal health for human posterity. Race poisonists embraced a radically more inclusive concept of paternal reproductive risk and responsibility than the notion of men's roles in reproductive outcomes that came to prevail later in the century. In the worldview of scientific eugenicists, a man's prenatal behavior was perceived to be at least as risky—and responsible—as a woman's for the well-being of future offspring.

In fact, for many race poison theorists, harm to men's sperm, compared to women's eggs, was of highest concern. Male gametes, they argued, are more exposed to environmental insults due to their presence in an external gonad. Men's overall behavior was also believed to carry a higher risk profile than women's.

Each sperm is "the bearer of an extraordinary specific potential energy," with "effects far more complex and wonderful than the emanations of a similar sized speck of radium."[58] So wrote British physician, psychiatrist, syphilis specialist, and eugenicist Frederick Walker Mott, famous for his invention of the theory of shell shock to explain men's long-lasting psychological trauma after the experience of wartime combat, in his 1914 *Nature and Nurture in Mental Development.*

In a manner characteristic of writings about male reproductive risk during this time, Mott argued that "it is hardly conceivable that the germ cells are uninfluenced by a continuous saturation of the blood by poisons."[59] In men with syphilis or who engage in excessive alcohol consumption, "the male germ cells, which are continually building up the spermatozoa out of constituents taken from the blood, may by analogy suffer in their specific energy and vitality." Through this "devitalizing agency," Mott theorized that "weakly types of offspring will be produced," and "carried on in several successive generations."[60]

Race poisonists stressed the importance of educating boys and men, above all, on eugenic principles. Considering the idea of eugenic motherhood "essentially natural to the normal girl," Saleeby warned that "it is the eugenic education of boys that is more difficult." "If alcohol and syphilis, for instance, can be demonstrated to be what I would call racial poisons, the young patriot must make himself aware of their relation to parenthood, and must act upon his knowledge of that relation."[61]

As these passages show, public health alarm in two areas—male alcohol use and the prevalence of venereal disease in men—drove the special concern eugenicists expressed for the health of male sperm in the early twentieth century. Temperance activists and eugenicists converged on a call for germ plasm hygiene, warning of the dangers of so-called drunken conceptions in which the parental state of inebriation at the time of intercourse led to impaired vitality of offspring.[62] Mott, for instance, referred to "alcoholic poisoning of the parents" as a "devitalizing cause" leading to long-term developmental and intergenerational harms.[63] Saleeby implored that "parenthood must be forbidden to the dipsomaniac, the chronic inebriate or the drunkard, whether male or female."[64]

While both male and female alcoholism was thought to contribute to birth defects and genetic degeneration, the overwhelming focus of eugenic writing on alcohol's harms was on paternal alcohol use.[65] Historically, writes medical sociologist Elizabeth Armstrong, "greater emphasis on the paternal role in the hereditary effects of alcohol . . . was a reflection of the primacy accorded to the male role in reproduction and society, as well as the greater social latitude permitted men, their consequently greater tendency to drink and be drunkards, and the social assumption that most drunkards were men."[66]

In the United States, temperance movement leaders—in large part, women—constructed the male alcoholic, who wastes his family's income at the bar, brings violence to his home, and transmits hereditary weakness and venereal disease to his wife and children, as the singular public enemy that temperance laws could eliminate. Alcohol should be banned, they argued, for the safety of women, children, and the race. Whereas in the nineteenth century temperance authors grounded such claims in vague biological mechanisms, accepting a combination of impressions theories and Lamarckism, in the early twentieth century activists reframed their pleas in terms of the protection of the genetic integrity of the germ plasm.[67]

While acknowledging that maternal alcohol use during pregnancy could also "poison" the fetus, eugenicists drew attention to evidence that paternal alcoholism alone could harm future offspring. They cited family studies that, they argued, showed intergenerational patterns of disability caused by paternal drinking alone. Forel, for instance, described one study "of ten large families in which the father and perhaps some of the other forebears were drinkers," leading to "57 children, of whom 12 died of weakness . . . 8 suffered from idiocy, 13 from convulsions and epilepsy, 2 were deaf and dumb, 5 suffered from inebriety with epilepsy or chorea, 3 from bodily deformities, and 5 were dwarfed."[68] One can imagine that such claims were received as alarming statistics indeed.

Experiments done with healthy mothers, Saleeby similarly emphasized, "prove that *paternal* alcoholism alone . . . can determine degeneration."[69] "If a man saturates his body with alcohol carried by his blood, he injures all the tissues which are nourished by that blood, including the racial elements of his body with the rest," wrote Saleeby.[70] "The importance of the demonstration as regards the father," he wrote, "is that it means a poisoning of the paternal germ-cell."[71] Echoing a rallying call of the temperance movement, Saleeby advised, "An excellent eugenic motto for a girl . . . is 'the lips that touch liquor shall never touch mine.'"[72]

In the early twentieth century, venereal diseases such as syphilis were widely appreciated as a major public health problem. In *Being Well Born*, Guyer claimed that half of all men carried a venereal disease at some point in their lives.[73] Venereal diseases were also increasingly recognized as an originating source of many brain and mental illnesses. While men were considered the principal vectors of venereal disease, married women and their children were regarded as its most grievous victims. One author, in 1904, estimated that an astonishing 80 percent of married men had gonorrhea and that 70 percent of married women with a venereal disease contracted it from their husband.[74]

Eugenicists recognized diseases such as syphilis in either parent as a

danger to the fetus. But when married women were found to have venereal disease, and to have transmitted it to their children, causing treacherous outcomes such as infantile blindness, eugenicists were firm that the
cause was not the woman's behavior, but her neglectful male partner who
introduced it to her. In *Being Well Born*, for instance, Guyer used the male
pronoun in scolding "the syphilitic" and encouraged women to "demand
a clean bill of health on the part of their prospective husbands." When
men "realize the danger their condition imposes on wife and children,"
they should be "eager to know their condition and to have medical advice."[75] To critics of restrictive marriage laws requiring men to be free of
venereal disease before obtaining a marriage license, Guyer responded:
"To those who hoist the flag of personal liberty, it may fairly be asked,
how much personal liberty does the syphilitic accord his doomed and
suffering wife and children?"[76]

Marshall's prescriptions for prophylactic treatment of syphilis also
principally offered guidelines for the father. "It is usually the father who
introduces syphilis into the family circle," Marshall began, "and we have
seen that treatment of the father alone may lead to the birth of a healthy
child after one or more syphilitic one[s]. . . . Hence the importance of
treatment of the father and the prospective father." Marshall's guidelines held that men should refrain from marriage and reproduction for
"at least three years" while undergoing treatment, until "he has been
without symptoms for at least a year." Even when symptom-free, Marshall believed that any man who had ever suffered from syphilis should
prospectively undergo six months of treatment before starting a family.[77]

Eugenicists supported a host of laws to prevent the transmission of
venereal disease from men to women and children, proposing, for example, that men who tested positive on the newly available Wasserman
reaction test for syphilis should be denied a marriage license. "No other
disease can rival syphilis in its hideous influence upon parenthood and
the future. But it is no crime for a man to marry, [and] infect his innocent
bride and their children," bemoaned Saleeby.[78]

The first laws requiring venereal disease testing prior to marriage,
installed in the early 1900s, applied only to men. The impetus behind
these laws was not only concern over infection of women, and via this
route, any offspring, but also future descendants. A significant debate
in early twentieth-century syphilology concerned the possibility of inter-

generational transmission of syphilis through the sperm itself. In his authoritative third edition of *Syphilology and Venereal Disease* (1914), British physician Charles Frederic Marshall insisted that through "spermatic infection" a man may "injur[e] posterity" "whilst not a symptom has ever been observed in his wife."[79] In such cases, referred to as "Conceptional Syphilis," a healthy mother could even be infected by a "foetus syphilized by spermatic infection from the father."[80] Marshall argued that transmission of syphilis through the sperm was more dangerous to the fetus than its acquisition during the intrauterine period, leading to more severe congenital abnormalities and derangements, ranging from stunted growth, late developing speech, dental abnormalities, hare-lip and cleft-palate, to dystrophies and deformities of all kinds, as well as increasing the chance of hereditary transmission to future generations.[81]

PRECONCEPTION ADVICE TO FATHERS, FROM PRENATAL CULTURE TO SCIENTIFIC EUGENICS

The premise that prenatal and preconception advice should be delivered to men as well as women does not itself represent a radical break from previous eras. As noted in chapter 3, nineteenth-century parental impressions theories included paternal influences prior to and at conception as an element of the total "prenatal" impression received by the offspring.[82] In *The Science of a New Life* (1869), for example, John Cowan had maintained that the ideal conditions for conception involve diligent preparation on the part of both mothers and fathers. Conceiving a well-born child requires a loving husband and wife "fixed on the qualities to be transmitted, and the date for conception" and at least "four weeks of preliminary preparation," including sleeping "in different beds, if not in different rooms," dressing appropriately, and praying, exercising, and breakfasting properly.[83]

Prenatal culturist and chastity activist Mary Teats warned that failure to abstain from sex could degrade men's sperm. Men who "squandered" their "life force" have sperm that "are languid, moving about slowly" with an "abortive" tail, lacking "vigorous" "power of locomotion."[84] "If a man sows the seed (spermatozoa) saturated with alcohol, tobacco or licentiousness," Teats claimed, he cannot "expect a clean, pure, strong harvest in the resultant child-life." Sperm, she argued, "takes on and is affected by the thoughts of the individual," which will "saturate, charge and surcharge the zoosperms with the characteristics of sensualism and selfishness."[85]

Despite their much greater emphasis on the maternal role in forming mental and character traits of children, prenatal culturists made frequent mention of the duties of the father at all stages. Referring to the sperm, Georgiana Bruce Kirby advised in 1889 that "the material supplied by the father . . . represents his then state of being, and will continue to represent it in the life it has helped to organize." "This finest secretion of the man's whole being—this subtle essence of his nature, which is both spiritual and physical—should express his best possible condition," she wrote. "After this," however, "he can only affect the child indirectly through his influence on the mother's mind."[86]

Prenatal culturists suggested that men cultivate their bodies and characters in advance of conceiving a child. Newton Riddell related the cautionary tale of a sedentary male friend whose son was born "sadly wanting in physical development and vitality." After the father "took up systematic physical culture" and became "a well-developed athlete," his next child shared his improved physique. In anticipation of conceiving a child, men should give up alcohol and tobacco, improve their personal cleanliness, exercise vigorously, and avoid all manner of unvirtuous be-havior: "The tendency toward loafing may be as fully transmitted as any other acquired character," averred Riddell.[87]

"Does the father, in preparing the germ, so impress on it his own conditions of body and soul that these must necessarily be developed in the future child, so as essentially to affect his character and destiny? That he does is certain," wrote William Truitt in his 1914 *Nature's Secrets Revealed*. "Whatever diseases affect the father must also affect the secre-tions of his system, and none more so than the germs of future human beings. What an obligation, then, rests on every man, to see to it, so far as he can, that the system in which the id-germs of human existence are prepared should be replete with manly beauty, tenderness and power!"[88]

Fathers were also directed to support their partners in providing the best maternal environment for the fetus during gestation.[89] Under the heading "The father's influence through the wife," Kirby related a story illustrating how a father might affect his future child's character during gestation. A "Mr. Z" read to his pregnant wife "many of the standard

English poets and essayists" and sure enough, the child had a superior mind and intellectual habits, despite the reputed dullness of the young wife.[90] Fathers should be wise and patient, continue their healthy habits, and provide a stable environment for the pregnant woman. A loving husband, prenatal culturists argued, can help the pregnant woman realize her highest gestatory aims. Wrote Blanche Eames, "In reality the mother wields a wonderful power, either for good or ill, over her unborn child. But the father's influence during this period must not be overlooked, since to such a large degree he may make or mar the happiness of his wife, thereby making possible or impossible a propitious environment for the child."[91] Emma Drake agreed, suggesting that male smokers produce pregnant wives "struggling along in ill health"; the infant, she suggested, similarly suffers from "antenatal and postnatal poisoning" from the father, and "as soon as it is old enough it will take up the habit which is already acquired, to pass down along the line a more and more enfeebled heritage."[92]

As these passages demonstrate, despite the mantle of modern science, early twentieth-century scientific eugenicists who expressed special concern about the eugenic integrity of male gametes extended longstanding traditions in their reproductive advice to men regarding the health and well-being of their sperm. First, lingering in eugenicists' prescriptions for men is the impressionists' intuition that one's total constitutional health leaves a potentially heritable trace on the gamete. Second, scientific eugenic discourse on the sperm often replicated the historical presumption that the sperm is the seed and seat of virility, while the mother's principal contribution is nutritive and gestational.

But scientific eugenicists' pronouncements on male germ plasm hygiene differed in key ways from the prenatal culturists. While prenatal culture theorists situated sperm health as one small component of shaping future offspring, they were firm in their view that mothers had a far greater influence, through gestation, than did fathers on the constitution and character of the child. Scientific eugenicists inverted these priorities, focusing on germ plasm hygiene while minimizing the importance of influences during gestation. In their emphasis on the threat of so-called race poisons to male sperm, scientific eugenicists also narrowed their target to specific chemical factors, departing in significant ways from the expansive picture of the hereditary imprint of paternal and maternal behavior given by prenatal culturists.

PRENATAL ADVICE TO WOMEN

Above all, scientific eugenicists' prenatal advice to women emphasized the need to take stock of a man's health before conceiving a child with him, so as to avoid the influence of a compromised sperm. A woman, Saleeby quipped, is "a practical eugenist [sic]" when she selects her husband.[93] Girls educated in eugenics should be "taught to look upon" behaviors such as alcohol use as "tabooed."[94] A "fast or dissipated young man" who is "unsound physically" "is an actual and serious danger to his future wife and children," wrote Guyer. Hence women must "see the necessity of demanding a clean bill of health on the part of their prospective mates."[95]

The notion that the health of a man's sperm should be a woman's concern can be found frequently in early feminist writings about birth control and female choice in marriage. A 1907 essay by British birth control advocate Montague Crackanthorpe, for example, offered just this sort of feminist eugenics argument: "The ideal woman of the twentieth century is a fully developed creature. . . . No longer limited to the roles of mistress, wife, mother, she has come to recognize the all-important part assigned to her in the destiny of the race. She realizes that on her depend its ascent, its descent, its very continuance. . . . She declines to give to her children for father the degenerate, the drunkard, the physically or mentally unfit."[96]

Moreover, eugenicists framed many risks to female germ plasm and intrauterine health as best addressed by modifying *men's* behavior. Eugenicists often encouraged men to marry on the basis that it encourages "the support of motherhood by fatherhood."[97] This included a father's presence and support during gestation, which was crucial, they argued, to his children's health. A chapter titled "A Man's Duty toward a Pregnant Wife" in the 1916 manual *Manhood and Marriage* advised that, "The differences in the physical vigor and constitutions of different children of the same parents may invariably be traced to differences in the conditions which prevailed previous to conception and during gestation. To take proper care of his wife at this time, and to do his utmost to provide for her physical and mental welfare, therefore, is not only a duty that a man owes to her, but it is a duty that he owes to his children as well."[98]

Certainly, eugenics-era physicians and scientists acknowledged the importance of maternal health during pregnancy to offspring outcomes. The 1913 US Children's Bureau pamphlet on prenatal care, for example, advised that during pregnancy a woman should "order her own life in the way that will result in the highest degree of health and happiness for herself and, therefore, for the child." A woman who "goes through her pregnancy pining or lamenting her condition," the pamphlet warned, may produce a "puny, wailing baby" with a "nervous condition." This is not due to maternal impressions, the pamphlet clarified, but because the mother's state left the child "robbed of the nutrition he needs."[99]

While scientific eugenic maternal health advice dismissed claims that the visual and emotional impressions of the mother could transfer directly and literally to the offspring, it also often advised that nutrition, trauma, high emotion, and the direct environment of the gestating mother could contribute to the birth of a more or less robust infant. Thus, despite their strenuous denunciation of maternal impressions as mere superstition, some advisors' prescriptions to pregnant women often sounded startlingly like those of the prenatal culturists. In the *Mothercraft Manual*, for instance, Read advised that "the mother is influencing the child during the entire nine months, through the blood supply. If she indulges in fear, anger, melancholy (dark emotions that develop poison in the blood), if she over-eats, or takes alcohol, if she neglects deep breathing, daily bathing, elimination, exercise, she is impoverishing and poisoning the blood supply, and the quality of her child's bodily and mental characteristics will suffer."[100] And *Facts about Motherhood* asserted that "the harm which a mother may pass to her fetal child is not the physical marking," a reference to maternal impressions, "but through improper ordering of her own life, she may disorder the child's future happiness by giving it an impaired constitution."[101]

John William Ballantyne, the Scottish obstetrician today recognized as the founder of prenatal care, took such rhetoric to Olympian heights in his 1914 *Expectant Motherhood: Its Supervision and Hygiene*. The mother's "body is [the infant's] immediate environment, and he is profoundly affected by it for good or evil." During pregnancy, the mother can "alter for weal or for woe the child as yet unborn." Ballantyne even suggested that "the regulation of the immediately pre-natal influences which play upon the child" might have "even greater effect than those which are termed hereditary" in shaping the infant.[102] In light of the mother's critical role in shaping the health of her offspring, Ballantyne echoed the prenatal culturists, advising that she should be protected from "tragic" and "gruesome" sights, and "misdemeanours of those she loves" should be kept secret from her: "a sort of beneficent conspiracy."[103]

Scientific eugenicists primarily concerned about the so-called race poisons also did not neglect intrauterine influences. "Any definition of eugenics," argued Saleeby, involves "selection for parenthood based upon the facts of heredity" as well as the edict to "take care of those selected—as, for instance, to protect the expectant mother from alcohol, lead, or syphilis."[104] Eugenic scientists recognized several diseases—including syphilis—that could be transmitted to a fetus while in the womb. Eugenicists, too, were among those who supported new laws to prohibit pregnant and breastfeeding women from working in factories where they might be exposed to hard labor or toxins.[105]

But for eugenic aims, it was germ plasm hygiene, rather than intrauterine factors, that ranked the highest priority. "Beautiful are the antenatal defenses," wrote Saleeby, summoning the image of "healthy and vigorous" babies born of "some pallid, half-starved, stunted mother in the slums."[106] For his part, Forel mentioned the "hygiene of pregnancy" exactly once in *The Hygiene of Nerves and Mind*. "Diseases, emotional excitements, nutritional disturbances, and everything else that injures the bodily health and especially the nervous life of the mother naturally have more or less of an indirect effect upon the life of the embryo," he offered. "Yet," Forel assured, "since the nervous system of the latter stands in no direct connection with that of the mother it is affected only indirectly."[107]

Focused largely on the prevention of infant mortality in the first months and years after birth, eugenic advice to women provided guidance on postnatal infant feeding and hygiene. "Babies of the slums, seen early, before ignorance and neglect have had their way with them, are physically vigorous and promising," wrote Saleeby.[108] "We find that, so far as ordinary physicians are concerned, the majority of human babies . . . are physically healthy, at birth."[109] The issue, as Saleeby and many of his colleagues saw it, was the high rate of infant mortality due to neglect, disease, and poor nutrition in early postnatal life. Hence, postnatal infant care, feeding, and hygiene formed the core of maternal and infant health sciences in the early twentieth century. By contrast, advice for pregnant mothers offered commonsense generalities, and received little more than passing mention.

While scientific eugenicists, then, recognized the possibility of intra-uterine influences, most attributed minimal significance to them compared to the importance of maintaining germ plasm health. Convinced as they were that only germ cell harm could lead to dysgenic intergenerational effects, discussions of prenatal influences principally concerned behaviors that could lead to changes in the genetic composition of individuals and populations.

"DON'T WRECK A GOOD GERM PLASM"

In the first quarter of the twentieth century, scientific eugenicists recruited a conception of the microscopic germ plasm as the seat of all human traits and the reservoir of human heredity to a new vision of genetic cultivation of individual and public health. "We should look upon the germ as a living thing," wrote Conklin. The "units of the germ cells," he claimed, determine all aspects of human traits: sex, "nervous temperament," predilection to disease, and even "genius."[110] Whereas an individual person "develops and dies in each generation; the germ-plasm is the continuous stream of living substance which connects all generations. The person . . . is merely the carrier of the germ-plasm, the mortal trustee of an immortal substance."[111] Simultaneously, early twentieth-century biological, obstetric, and lay advice to women broadcast a uniform mantra: pregnant women should reject ideas about maternal prenatal impressions as superstition.

This different way of seeing reproductive risk reminds us how our scientific theories carry implications for the range of imagined interventions to protect and optimize offspring health. When the sperm, rather than the fetus, is elevated as the most plastic and vulnerable link in the reproductive transmission of hereditary traits, the gender politics of reproductive risk look quite different. Upon examination of scientific eugenics literature from 1900 to 1935, a distinctive historical moment in the history of perceptions of prenatal reproductive risk emerges: one of moderate advice to mothers and heightened concern about fathers.[112]

Contemplating this earlier conformation of the gender and sexual politics of reproductive risk, focused on germ plasm hygiene, challenges the assumption that it is simply common sense that the pregnant mother constitutes the most proximate and important point of intervention, control, and agency in managing reproductive risks. Eugenic-era prenatal advice to mothers and fathers also makes starkly clear the historical en-meshment of reproductive and genetic science in the politics of race, class, gender, disability, and nation. While today eugenics is recognized as founded on noxious racism and poor science, at the time, scientific eugenicists understood themselves to be participating in a progressive project of health uplift informed by the new sciences of genetics and the then-coalescing fields of obstetrics, pediatrics, nutrition science, toxicol-ogy, infectious disease, and public health.

In subsequent years, as interest swung back to environmental factors in prenatal development, the maternal body, constituted as the primary environment for the growing fetus, once again moved to the center of attention. The next two chapters tell the story of the emergence, in the mid-twentieth century, of a new framework for the scientific study of human maternal effects. Our first move, however, will be to go back in time, to some anomalous findings of maternal influences in heredity already apparent in the early days of Mendelian genetics.

CHAPTER 5

Maternal Effects

"Maternal effects" is a technical term in the fields of genetics, breeding, and developmental biology. First coined in a 1935 scientific paper on fruit fly genetics, the enigmatic term is widely misunderstood and often misapplied. Tracking the history of the concept of maternal effects from the advent of Mendelian genetics to the mid-twentieth century, this chapter reconstructs the debates and contexts in which it came to displace older notions of the maternal imprint.

Our journey, like the history of genetics itself, will take us first through plant genetics and laboratory studies of nonhuman animals like snails and fruit flies, before arriving at mammalian and human studies. We begin in 1905, in the outlying greenhouses of agricultural research stations, where researchers hoping to optimize the products of breeding were working to map Mendelian traits across a broad range of species. The newly rediscovered science of Mendelian genetics offered the radical promise of an unprecedented level of control over the products of reproduction. Discrete traits could be isolated and their hereditary patterns mathematically predicted. Through strategic breeding undesirable characters could be avoided, desirable ones cultivated, and novel forms created.

EXCEPTIONAL AND UNCONFORMABLE PHENOMENA

In 1905, Rowland Biffen, a young British botanist, began applying the new science of Mendelian genetics to experiments for the "improvement of English-grown wheat."[1] His research quickly gained notice for upending a fixture of pre-Mendelian agricultural breeding lore. Theories of so-called maternal prepotency, grounded in the Aristotelian notion that the womb forms the offspring as if it were clay, had long held that a hybrid offspring of two strains would most resemble the mother.[2] By contrast, Mendelism predicts that whether the male or female parent carries the trait, the pattern of inheritance will be identical. Biffen performed reciprocal crosses (mating a female of strain A with a male of strain B, and a female of strain B with a male of strain A) for dozens of wheat traits previously understood as driven by maternal prepotency. In a striking reversal of lore, the pattern of inheritance was identical whether the male or female parent carried the trait.

But in one instance, Biffen found a stubborn exception—the first in a series of disconcerting wrinkles reported in reciprocal cross studies by Mendelian geneticists. Crossing long- and short-grain wheat, Biffen discovered that "the shapes and colours" of the "hybrid grains of the reciprocal crosses . . . corresponded with those of the female parents." Acknowledging this inexplicable result, Biffen was compelled to conclude that "it seems clear that the maternal plant characters—in this case of size of the glumes—in some way directly influence the seed characters of each generation."[3]

The struggle to reconcile Biffen's finding with the framework of Mendelian genetics would last a quarter-century, until the introduction of the concept of maternal effects by Theodosius Dobzhansky. William Bateson took up the question of the "maternal impress"—illustrated by an image of Biffen's long- and short-glumed wheat—in 1909, in his classic laboratory and teaching manual *Mendel's Principles of Heredity*. "Of the various cases alleged as exceptional, or declared to be incompatible with Mendelian principles, few have any authenticity. Several rest on errors of observation or interpretation and some have even been created by a mistranslation or misprint," Bateson began, under the heading "Miscellaneous exceptional and unconformable phenomena." "The progress of research has gone steadily to show that the facts of heredity which at first seemed hopelessly complicated can be represented in terms of a strict Mendelian system." Exceptions to Mendelian rules, Bateson declared, "are so rare that announcements of discoveries irreconcilable with the principle of factorial composition may be safely disregarded, unless they are made by observers experienced in this class of investigation, or supported by evidence of exceptional strength."[4] But Bateson singled out Biffen's "curious and at present quite unintelligible phenomenon" of a maternal influence in heredity as one finding exceptional enough for special consideration.[5] This "peculiarity," Bateson wrote, cannot be simply dismissed as an error. But the causal mechanism was "recondite."[6] "How this influence is exerted we cannot suggest" as it "operates in an extraordinary capricious . . . manner."[7]

Bateson intuited that the concept of "inheritance" was not quite appropriate to describe the phenomenon, employing phrases such as maternal "impress" and "influence" instead. Biffen's long-grained wheat, Bateson wrote, presents "the remarkable fact that the mother-plant can impress varietal characters on her offspring *by influences which are not hereditary in the ordinary sense.*" Some presume that such influences, if real, are "negligible" in importance, but, Bateson speculated that "without any serious stretch of imagination we may suppose that a maternal impress may be such as to produce an effect lasting at least for the lifetime of the immediate offspring."[8] With this statement, the intuition that the maternal element confers something important to the offspring distinct from that of the paternal, and that this is an elusive force quite independent of the normal processes of hereditary transmission, was carried forward into the genetic age, seeding a new round of investigations and rekindling the longstanding debate about sex equality in heredity, thought to be extinguished once and for all by Weismann.

Other findings of apparent maternal influences in heredity soon surfaced. Japanese scientists investigating silkworm genetics reported a stream of "maternal inheritance" exceptions to Mendelian predictions, which became known in the English-speaking world beginning with a 1913 paper in the *Journal of Genetics* by Kametaro Toyama.[9] Toyama was studying how silkworm characters such as color—a prime concern for the Japanese silk industry—could be predicted and manipulated by following Mendelian laws. For one element of color, Toyama found that "there is a departure from the normal rules."[10] He observed that "there is no sign of male characteristics to be seen in the egg," and concluded that "the colour-characteristics of the egg are governed by those of the female parent."[11] Toyama hypothesized that silkworm egg color is determined by the character of the maternal egg prior to fertilization and is therefore outside of the influence of paternal genetics.[12]

In another Japanese contribution, in 1918, Hajime Terao examined the apparent maternal inheritance of seed coat color in the embryonic soybean. In reciprocal crosses, offspring seed coats replicated the color— yellow or green—of the female parent. "Hence we are probably dealing with characters which can be inherited only through the female parents," reasoned Terao.[13]

British scientists made similar discoveries about maternal inheritance in the context of another animal: snails. In 1923, University College London physician Arthur Edwin Boycott and the amateur malacologist Captain Cyril Diver reported a series of findings on the apparent "maternal inheritance" of shell coiling in snails.[14] Snail shells usually coil in a right-handed spiral. Such spiraling is a simple visual trait with systemic consequences for the snail. It begins in very early development and refracts through the snail's whole body plan: a snail with a right-handed spiraling shell has a heart, kidney, rectum, and nervous system ordered to match it. When, rarely, the shell coils to the left, the snail's body plan is, as Boycott and Diver put it, "a complete mirror image."[15]

To learn about how this trait is inherited, the researchers bred the snails in green jam jars with loose glass stoppers on the roof of the medical school in Central London. In the summer, they constructed shelters to keep the snails out of the heat. In the winter, they brought them inside. Through years of reciprocal breeding and careful tracking of the ratio of offspring shell spirals, Boycott and Diver found that the offspring of female left-coiling—or sinistral—snails are always also sinistral, whereas the offspring of male sinistral snails mated with female right-coiling snails are not. This trait, they argued, is not genetically inherited but conveyed by maternal phenotype alone (fig. 5.1).[16]

In a book-length write-up of these studies, Boycott and Diver offered the first extended theoretical synthesis and discussion of maternal-inheritance exceptions to Mendelian inheritance. After a detailed description of their snail experiments, they ventured by way of conclusion that, across the natural world, maternal phenotype may frequently shape

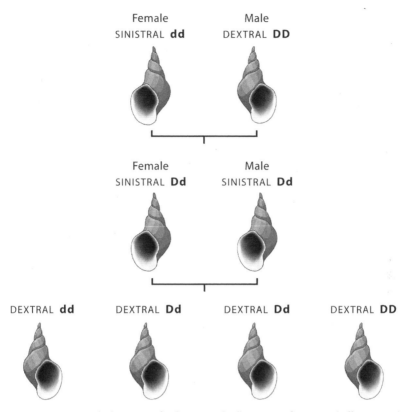

FIGURE 5.1 Maternal inheritance of coiling in snails. Illustration of a maternal effect on snail shell coiling in the first generation. Results in the second generation show a typical Mendelian pattern of inheritance. From Gurdon, "Sinistral Snails and Gentlemen Scientists," based on Boycott and Diver, "On the Inheritance of Sinistrality in *Limnaea peregra*." By permission of Elsevier.

early-forming traits. Perhaps, they suggested, "every embryo starts by being maternal."[17]

These varied studies prompted new theoretical interest in maternal influences in heredity among geneticists. Did maternal influences pose a devastating challenge to the predictions of the Mendelian laws of inheritance?

MATERNAL INHERITANCE

In their early studies of maternal influences in soy and silkworm, Japanese researchers had used the term "maternal inheritance" to describe their observations, engendering consternation. Geneticists invested in locating heredity purely in the genes were leery of describing observations of the effects of the maternal phenotype on the early developmental character of an embryo as, strictly speaking, a form of "inheritance." But even if the term "inheritance" were to be accepted, the modifier "maternal" certainly could not.

Thomas Hunt Morgan, architect of classical genetics, addressed the matter of "'peculiarities' determined by the character of the mother" in his 1919 *The Physical Basis of Heredity*.[18] Most cases of putative "maternal inheritance," he argued, "do not differ in any essential respect from ordinary heredity."[19] As he reasoned, in reference to Biffen's classic study, a long-grain "female" wheat mated with a short-grain male will have all long-grain offspring (the F_1 generation). But the sperm will influence the offspring's gametes, and in the next generation, the hereditary influence of the sperm will be fully present. That is, the gametes of the offspring (which form the F_2 generation) will have typical Mendelian ratios—thus, the effect is "deferred" by a generation, as illustrated in fig. 5.1. Morgan proposed that the term "deferred nuclear influence," rather than maternal inheritance, be used for such instances.[20]

Writing in 1923, Japanese geneticist Hajime Uda echoed Morgan's critique, objecting to "very ambiguous" uses of the term maternal inheritance. "So-called maternal inheritance," wrote Uda, is only "apparently [in] conflict with Mendelism." "It would be hasty to conclude immediately from these facts that a new type of heredity is involved." The sperm's effect is simply delayed—hence, he proposed that rather than calling this form of inheritance "maternal," it should be referred to as "delayed."[21]

Timing was key to the distinction Uda posed between what he called the "maternal impress" and true inheritance. Only characters that emerge very early in embryonic formation are influenced by the mother in this way, he argued. In botany, "no example is yet known in which the maternal impress extends beyond the seed stage." In such cases, Uda reasoned, "the male gamete, being introduced into the egg at the moment of the fertilization, has not time enough to manifest its own effect even if it be dominant." Uda thus strenuously argued that there is no need to recognize "a new form of inheritance" or "a special kind of heredity" to accommodate apparent cases of maternal inheritance because the effects appeared too early in development to count.[22]

Morgan's collaborator and prize student A. H. Sturtevant embraced Uda's view in a 1923 commentary in *Science* on the case of apparent maternal inheritance of left-spiraling in snails. So-called maternal inheritance was, he argued, a nuclear chromosomal effect imprinted in the egg prior to fertilization and the activation of the sperm. Discussing Boycott and Diver's findings, he argued that the inheritance of left-spiraling shells is "an exceptionally clear illustration of 'maternal' inheritance that is nevertheless dependent upon the chromosomes." The "case is a simple Mendelian one," in which the effect of the Mendelian characters carried by the sperm "is delayed for one generation." Hence, the maternal influence on snail coiling was both "a model case of the Mendelian inheritance of an extremely 'fundamental' character" even as it is also "a character that is impressed on the egg by the mother."[23]

Yoshimaro Tanaka made similar arguments in 1924. Reviewing reports of supposed "maternal inheritance" in sea urchins, fish, snails, corn, soy beans, and silkworms in the journal *Genetics*, Tanaka argued that such cases "should not be called maternal in inheritance," per se, because the inherited traits, such as egg shell colors, "are produced in the maternal body before fertilization, and therefore belong to a different generation from the embryo contained in the same egg."[24]

When Boycott and Diver published their manuscript-length précis of studies of the inheritance of sinistral shells in snails in 1930 they no longer used the term "maternal inheritance" to describe the hereditary pattern that they had uncovered. "The term 'maternal inheritance,'" they wrote in justification, "has been used in a loose and irregular way to cover a variety of phenomena."[25] Boycott and Diver hence chose to use the phrase "delayed inheritance," over several other competing prospects, including "pre-inheritance," "deferred nuclear action," and "accelerated character determination."[26] The explicit aim was to ensure consistency with the principle of sex equality in heredity. "Whichever expression is used," they wrote, "it would be well to restrict it, as we do here, to a form of inheritance in which both sexes contributed factors in the normal manner."[27]

Despite such mental gymnastics to ensure that language did not challenge Mendelian principles, by the 1930s, accumulated reports of apparent maternal-line exceptions to Mendelian laws in a variety of species led some geneticists to gradually accept that a set of varied phenomena, unconformable, in their full scope, to Mendelian laws, operated as an elusive force all of its own in heredity. But just what to call these exceptional and unconformable phenomena?

MATERNAL EFFECTS

The problem, as Bateson had put it, was that if maternal influences are in any way to be understood as a component of variance in heredity, they are surely not heredity in the "ordinary sense." In 1934, the eminent German geneticist Richard Goldschmidt took up the matter.[28] Goldschmidt was a big-picture conceptual thinker who, over his career traversing the Kaiser Wilhelm Institute for Biology in Germany and the University of California at Berkeley, argued for a conception of heredity beyond that of simple genetic transmission. For Goldschmidt, maternal influences presented a perfect example of the kinds of phenomena he felt were excluded from reigning genetic theory.

Goldschmidt's longstanding interest in maternal influences stemmed from studies of "large crosses" hybridizing parents at a great genetic distance from one another. In his own work, Goldschmidt performed reciprocal crosses among distantly related strains of caterpillar. That is, he bred a male of type A with a female of type B, and the opposite. The offspring, he argued, showed consistent "direction towards the maternal type."[29] In a review of the literature, he documented findings across many species of plants and animals suggesting that "certain hereditary traits . . . especially the growth-habitus" show maternal influence.[30]

Goldschmidt referred to maternal influences in heredity as "cytoplasmic influence upon hereditary characters," but context makes it is clear that Goldschmidt was not speaking only of cytoplasmic inheritance, per se, but of a broader category of maternal imprints, for which the mechanism was still poorly understood. Not to be confused with the "germ plasm," which refers to the material inside the nucleus, cytoplasm is the jellylike material inside of the cell membrane but outside of the nucleus, which can be transmitted to offspring via the maternal egg. Cytoplasmic inheritance, the transmission of genetic material to offspring from the cytoplasm of the maternal egg, was often conflated with other forms of putatively "maternal" influence in heredity during this period.

Prior to Weismann's nuclear germ plasm theory of heredity, which became widely accepted in the 1890s, leading embryologists believed that heredity involved the passage of both nuclear and cytoplasmic material to the offspring. Cytoplasm was also considered a critical chemical milieu for metabolic processes involved in growth and development. In the late nineteenth century, wide consensus held that the offspring, at least in some small respect, receives its earliest developmental instructions from the maternal parent through the egg cytoplasm, and some of the most sophisticated scientific challenges to Weismann's germ plasm theory of heredity would come from partisans of cytoplasmic inheritance.

By the turn of the twentieth century, the broad strokes of Weismann's view were settled among scientists, and cytoplasmic inheritance was decidedly out of fashion. However, even as geneticists such as Thomas Hunt Morgan were unraveling the precise correlations between the chromosomes and Mendelian traits, some continued to insist that the maternally inherited cytoplasm, not the nuclear genes, carried the basic plan of growth and development that characterizes the species. The German-American chemist Jacques Loeb, for example, remained an outspoken defender of the cytoplasmic theory well into the second decade of the twentieth century. Loeb's 1916 *The Organism as a Whole* asserted that "the cytoplasm in the egg is already the embryo in the rough."[31] Harkening to Brooks, Geddes, and Thomson, and still further, back to the seventeenth-century ovists (see chapter 2), Loeb claimed that "Mendelian factors in the chromosomes can impress only individual characteristics," but that the overall species-type of the organism is determined by the egg.[32]

But most scientists regarded Loeb's views as confused and outdated. In his classic 1896 synthesis *The Cell in Development and Inheritance*, the American biologist Edmund Wilson summarized the case against cytoplasmic theories and in favor of the germ plasm theory. Studies of single-celled organisms demonstrated that regeneration "can only take place when the nuclear matter is present."[33] Additionally, they showed that all of the mitotic division of chromatin occurred within the nucleus of the cell, which is "equally derived from both sexes." It is this nucleus that, in turn, produces the egg cytoplasm itself.[34] Simply put, the nucleus "is, so to speak, the ultimate court of appeal."[35]

Some discrete, well-documented forms of cytoplasmic inheritance were accepted by Weismannians. The phenomenon of non-nuclear DNA inheritance in mammals in the form of mitochondrial DNA transmitted in the egg cytoplasm was not discovered until 1963. But already by 1909, geneticists had reported that in many plants, elements are passed through the maternal cytoplasm that influence chlorophyll functioning, and hence pigmentation, in the resulting offspring.[36] Thus, at first, Japanese soybean geneticists had speculated that observations of maternal influences in the inheritance of coloring might be attributed to "inheritance of chlorophyll through cytoplasm" (a factor later ruled out).[37]

To be sure, Goldschmidt was no Loeb, fighting for the cytoplasm against the nuclear monopolists.[38] He agreed with Morgan and other contemporaries that the cytoplasm is not "the seat of definite hereditary properties in the same sense as this is true for the nucleus."[39] Furthermore, he accepted the concept of deferred or delayed heredity as fully explanatory of many cases of so-called maternal inheritance. Toyama's "much-quoted maternal inheritance," he conceded, "turned out to be in fact chromosomal."[40]

Rather, Goldschmidt thought of "cytoplasm" in a much broader sense, as the totality of the maternal phenotypic contribution to the embryo. This maternal "substratum" or "background," he hypothesized, could alter "the action of the genes in controlling hereditary traits," so that the phenotype in many reciprocal crosses resembles the mother, in ways more gestalt than discrete: "size of organs or parts, amount of pigment, shapes of parts and similar characters."[41]

This form of heredity, Goldschmidt stressed, does not operate in a manner akin to standard Mendelian traits under genetic control. In his telling, maternal influences are a milieu, background, or substratum that acts via "velocities," "inhibition," and "developmental reactions." Maternal influences of this nature, Goldschmidt hypothesized in an intriguing nod to longstanding ideas linking the maternal imprint and intergenerational influences of the environment on heredity, may provide a means for "external conditions" to alter "certain characters of an organism and the change reappears in the next generation in a lesser degree and might continue thus for a number of generations until it finally disappears."[42]

The conceptual intermingling, apparent in Goldschmidt's discussion, between cytoplasmic inheritance and what many seemed to recognize was a different order of maternal influences in heredity altogether, reflected the sporadic, marginal, and roughly hewn state of scientific discussions of the question during the first decades of the twentieth century. This state of affairs was at last broken in 1935, with Russian-American geneticist and evolutionary theorist Theodosius Dobzhansky's introduction of the disambiguating term "maternal effect."[43]

The first use of the term was almost an aside. The immediate object of Dobzhansky's attention, in the short article appearing in the journal *Science*, was a finding of a difference between reciprocal crosses of a species of fruit fly, showing maternal influence on testes size in male offspring. Dobzhansky defended this as a rare example of a true case of maternal influence, explained not by factors in the cytoplasm but by the makeup of the mother from whom the cytoplasm was derived. "The cause of the difference between reciprocal crosses has not been adequately analyzed in most cases," Dobzhansky wrote. "It is a 'fallacy' that such findings are automatically seen as evidence for 'cytoplasmatic [sic] inheritance.'"[44] In the case of the fruit fly testis, Dobzhansky wrote, "We are dealing . . . with a *maternal effect*," meaning that the character of the embryo was determined exclusively by the maternal phenotype, not by genes or cytoplasmic elements transmitted from parents to offspring.[45]

With the term "maternal effect," Dobzhansky defined a conceptual category, separate from maternal or cytoplasmic inheritance, per se, for maternal tendencies in heredity. Dobzhansky's term "maternal effect," disposing of "inheritance" in favor of the phenomenological term "effect," left unspecified the *cause* of the effect, and remained vague as to how to locate the effect within heredity. This fuzziness became its virtue. "Maternal effects" became an open-ended explanatory horizon for maternal influences that fell outside of the realm of cytoplasmic transmission and were unaccountable by any standard genetic mechanism. The term continues to beguile to this day.

The concept of "maternal effects" as conceived by Dobzhansky in the 1930s, however, extended only to phenotypic effects of the female parent on the early embryonic development of plants, insects, and other invertebrates. It had not yet been postulated in mammals—that class of animal, of which humans are a part, that incubates its young for sometimes quite extended periods in the mother's womb. But it was about to be.

THE SHETLAND AND THE SHIRE

The average Shetland pony, originally bred in the Shetland Isles of Great Britain, stands less than three and a half feet tall. In contrast, the British work horse known as the Shire is the largest horse in the world, at about six feet in height. Shortly before the start of World War II, in 1934, genetic researchers at the University of Cambridge Animal Research Station undertook a reciprocal cross experiment—artificially inseminating a female Shetland with the sperm of a Shire, and a female Shire with that of a Shetland—"to see how far the size of the mother would affect the size of the offspring."[46] Would a Shire give birth to a typically sized Shire foal, something halfway between the two, or something closer to a Shetland foal?

Cambridge professor John Hammond directed the experiment. Today, Hammond, author of *Farm Animals: Their Breeding, Growth, and Inheritance*, is remembered as "the father of modern animal physiology," one of the key developers of the "artificial vagina"—still today a mainline technology for extracting sperm for artificial livestock insemination—and a "guru" of the twentieth-century livestock sciences.[47] Although the Shetland-Shire experiment may at first glance seem a bit of a sideshow folly in comparison to today's sophisticated animal laboratory research, its aims went to the heart of going scientific questions in genetics and animal breeding. The goal, wrote Hammond, was to assess the "relative importance of heredity and environment in determining the size, development, and ultimate proportions of animals." Hammond reasoned that if the size of the resulting offspring was closer to the maternal type, the difference "would therefore be due not to chromosomal differences but to differences in the environment brought about by difference in the size of the mother. In other words we would have a controlled experiment in which 'Mother-size' was the only or predominating variable."[48]

By the 1930s, when Hammond conducted this study, inheritance of body size had already been well established as a classic problem in studies of the interaction between heredity and environment. Scientists across the fields of genetics, embryology, agricultural research, nutrition science, and anthropology had engaged the question in the first decades of the twentieth century. At the time of Hammond's experiment, the most influential body of work came from geneticist William Castle, who through decades of experimentation on size inheritance in rabbits, guinea pigs, and mice at Harvard's Animal Research station, the Bussey Institute, argued that mammalian body size was a completely genetic trait and was most often halfway between that of the male and female parents.[49] Hammond's results, however, gave support to the opposing view: that maternal intrauterine effects influence body size.

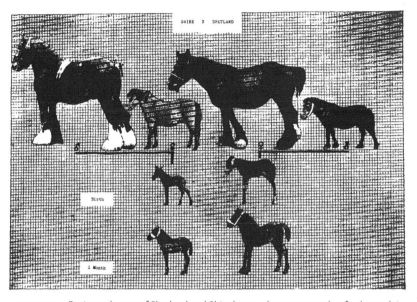

FIGURE 5.2 Reciprocal cross of Shetland and Shire horses demonstrates that fetal growth is constrained by maternal size. From Walton and Hammond, "The Maternal Effects on Growth and Conformation in Shire Horse-Shetland Pony Crosses."

In a 1938 report published in *Proceedings of the Royal Society of London*, Hammond reprinted photographic cut-outs of the resulting foals arrayed against graph paper (fig. 5.2). The figure made the striking results clear at a single glance: a female Shetland mated with a Shire gave birth to a foal of very similar weight and dimension to a purebred Shetland; conversely, the female Shire's offspring was closer in size to a normal Shire foal. Concluded Hammond, "In each reciprocal cross there has been very complete regulation of the size of the foal, the foals in the small mothers being limited in size to that of Shetland offspring and those in the large mother growing unchecked, although perhaps not quite reaching the full size of a Shire foal." Moreover, as the foals grew into adults, their size differential persisted.[50]

The experiment demonstrated, Hammond claimed, that "it is the uterine environment rather than factors inherent in the fetus which determines" body size. "Maternal regulation," Hammond wrote, "has masked the genetical differences." Though Hammond did not cite Dobzhansky, he applied the term "maternal effect" to his finding.[51]

Hammond recognized that maternal intrauterine effects carried broad significance for debates over the relative importance of heredity and environment, as well as practical implications for breeding agricultural stock for size. An open question remained: "If," via maternal effects, "permanent differences in adult life were established, would these differences be transmitted to subsequent generations?"[52] Alas, the question went unanswered, as Hammond was forced to destroy the research animals during the war. It would not be long, however, before new methods in reproductive biology raised it again.

EGG TRANSPLANTATION AND THE SEARCH FOR "TRUE" INTRAUTERINE MATERNAL EFFECTS

Striking though the findings of Hammond's tiny study were, Hammond could not distinguish between a cytoplasmic effect—inheritance of some factor in the mother's egg cytoplasm that controlled growth—and a true effect of maternal uterine size alone. To demonstrate intrauterine maternal effects, a reciprocal cross is not sufficient. Uterine and cytoplasmic contributors can be distinguished only if one also studies the outcome when fertilized eggs of a cross-breed are transplanted and grown in a female of the paternal strain, allowing the assessment of cytoplasmic influences independent of intrauterine ones. In short, what is required is an elaborate egg transplantation experiment involving cross-fostering of eggs.

In Hammond's day, egg transplantation was out of reach. While successful egg transplantation had been reported in a mouse as early as 1935, the method carried a high rate of failure.[53] The finicky technique could certainly not be applied reliably enough for scientific experimentation. The barrier was broken in the late 1940s by a Swedish geneticist.

Ole Venge, of the Institute of Animal Breeding in Uppsala, an associate of the eminent agricultural scientist Ivar Johansson, succeeded in carrying out reciprocal crosses involving 142 litters of rabbits and 249 egg transplantations.[54] The meticulous study involved crossing rabbit strains of different sizes and—to dramatize the effect—colors, and then performing the egg transplantation procedure. To carry out a reciprocal cross and cross-fostering via egg transplantation, Venge began with four live rabbits. First, the egg donor was mated with a male of the contrasting strain. Twenty-four hours after mating, the donor was killed, fallopian tubes extracted, and eggs washed out and stored in a salt

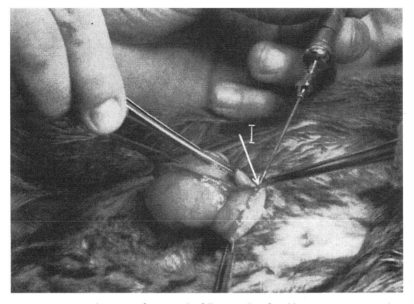

FIGURE 5.3 Transplantation of ova into the fallopian tube of a rabbit. From Venge, "Studies of the Maternal Influence on the Birth Weight in Rabbits."

solution (fig. 5.3). Meanwhile, the egg recipient was falsely mated with a vasectomized male, then anesthetized. A three-centimeter incision was made in the abdomen and eggs were then injected in the fallopian tube. The rabbit was sutured up, and Venge prayed that the eggs yielded a live birth (transplantations led to litters only a bit over 50 percent of the time).

The results—showing that a fertilized egg residing in a jet black Giant Flemish rabbit uterus grew 33 percent larger than in a smaller white Polish strain purely due to maternal effects—offered persuasive evidence for Hammond's hypothesis. Venge showed that the maternally provided intrauterine environment significantly modifies birth size, rendering the previous idea that mammalian body size was just a quantitative blending of genetic factors from mother and father impossible to sustain. Venge's 148-page study, published in 1950, was translated into three languages and became a citational classic in the field of maternal effects research.

Simultaneously, maternal intrauterine effects were under investigation in North America, in another context: the production of standardized rodents for laboratory research. Jackson Laboratories (JAX), founded in 1929 in Bar Harbor, Maine, today supplies more than

95 percent of the mice used in scientific research—close to two million animals a year.[55] The mice, the product of inbred strains developed to model human diseases such as cancer, are stored as frozen embryos, preserved in liquid nitrogen, and FedExed every day to research labs all over the world.

Today, lab mice are synonymous with modern biomedical science. But until the 1930s, mouse research was rare. Most geneticists worked with simple sea organisms, worms, and insects. Research on mammals largely occurred in the context of low-status "applied" agricultural research.

The creation of standardized inbred strains of mice intended for medical research, and the large-scale production and distribution of mammalian embryos, started in the 1930s and 1940s at JAX following huge infusions of funding created by the US National Cancer Institute Act. Crucial to the mass production of research mice was finessing the processes of embryo extraction and transplantation. The first successful transplantation, in 1935, was performed by Elizabeth Fekete in a mouse at JAX, though as noted above it would be at least a decade before such procedures were mastered.[56]

A key question for JAX scientists seeking to produce reliable mouse models for scientific research was whether embryos could be transplanted and grown in any uterus without affecting the characters of interest to researchers. Might some strains of mouse mothers produce bigger, more robust baby mice? How might the age of the mother, its nutritional state, its strain, and its size affect the litter? JAX researchers began examining these questions in order to address pragmatic aims, but they also recognized that the question of maternal intrauterine effects raised basic research questions in reproductive physiology, genetics, and endocrinology. In fact, JAX's studies of maternal effects in mice became one of the primary ways in which the laboratory portrayed itself to government and charitable funders as a powerhouse of basic research and not merely a mouse factory.[57]

JAX researchers, however, ran up against technological limitations. As early as the 1930s, JAX scientists had shown evidence of maternal effects in reciprocally crossed mice bred for high and low breast cancer susceptibility. At first, they believed the effect was intrauterine. But further research revealed that it was postnatal: low-cancer mice pups nursed by high-cancer mothers had higher rates of cancerous tumors.[58]

In 1942, Fekete provided suggestive evidence of a truly intrauterine effect. She transferred fertilized eggs of a black mouse into a white mouse, and vice versa. Fekete showed that the "uterine environment" of the black mouse was "more favorable for the development of eggs to living young" than that of the white mice.[59] The measured outcome of embryo viability, however, was too crude to suggest particular mechanisms.

A more striking finding came in 1951. When JAX scientists reciprocally crossed the black mice with another inbred strain, they found that offspring resembled the skeletal type of the female parent. "A difference between the reciprocal hybrids of inbred strains, present at birth . . . suggests that some features of the embryonic environment or of the cytoplasm of the ova must be different in the two strains," the researchers argued.[60] They acknowledged, however, that cross-fostering egg transplantation experiments would be required to "help to separate the effects of cytoplasm from the effects of uterine environment."[61] That confirmatory experiment, in mice, would not come for another seven years, across the Atlantic, in the London laboratory of Anne McLaren.

McLaren was a mouse expert and reproductive biologist. Today she is widely recognized as the mother of mammalian embryo transplantation. Her research helped pave the way for in vitro fertilization (IVF), a procedure that involves manually fertilizing an egg in the laboratory and then implanting it in the uterus. IVF has contributed to the birth of one million human babies in the United States alone since its first clinical application in 1978, and today is a $10 billion industry globally. A developmental biologist by training, McLaren dedicated her early research to the interaction of genes and environment in the mouse, leading her to the question of intrauterine effects.

In the 1950s, with the support of a grant from Britain's Agricultural Research Council, McLaren and Donald Michie, then her husband, engaged a series of targeted studies of what they called "uterine environment effects" in mice. A promising candidate lay in a finding, similar to that of the JAX scientists, that when reciprocally crossing two strains of mice, one with a yellow coat (agouti) and the other black, the offspring resembled the mother in a very specific way: number of vertebrae. If this could be shown to be a uterine effect, it would go beyond the findings of Venge and others who found maternal effects on "overall growth" to show "a specific morphogenetic modification, concerning a detail of form."[62]

The experiment would require 197 egg transfers over a period of 27 months, using three different strains of mice. "Experimental discrimination" between a cytoplasmic and uterine effect, McLaren and Michie wrote, "can be achieved by the transfer of eggs from one parental strain to uterine foster-mothers of the other strain. If the transfer has no influence on the character under study, then the egg cytoplasm must be responsible for the maternal effect; if it has an influence, the uterine environment is implicated."[63] First, McLaren and Michie performed a reciprocal cross of two strains of mice, one with five lumbar vertebrae, and one with more typical six. The offspring vertebrae generally resembled the maternal type in this experiment. Then, to discriminate between cytoplasmic and uterine influences, they took the fertilized eggs produced from the reciprocal cross, who had five vertebrae, and transferred them into a foster mother with six (fig. 5.4). If the effect was cytoplasmic, the offspring should have five vertebrae. But they didn't—they had six, like their foster mother. Last, a reverse transfer was performed, placing the fertilized eggs with six vertebrae into a foster mother with five. The offspring matched the foster mother's vertebrae count. The results showed incontrovertibly that "the uterine environment . . . exerts a specific morphogenetic influence upon the unborn young."[64]

Evidence that uterine environment alone can influence even the number of lumbar vertebrae of a developing mouse sent McLaren and Michie into excited speculation. First and foremost were the implications for genetic theory: "We must be prepared to extend our picture of the genetic control of morphogenetic processes," they wrote, to include "their regulation not only by the action of the embryo's own genes but also by the action of the genes of the maternal organism in which the embryo is gestated." But the implications were practical as well.

Maternal intrauterine effects, McLaren and Michie predicted, will become "an inevitable focus of clinical interest." Many human abnormalities may be traceable to "variables of the maternal physiology"—"some of them socially conditioned," they intriguingly suggested. And in agricultural breeding, egg transfer might become a major component of livestock breeding. Imagine, for example, the eggs of "outstanding females" transplanted to "genetically inferior recipients" for propagation of "cheap inter-continental transport." "It will then be of acute importance to know whether the offspring of such transfers will remain entirely

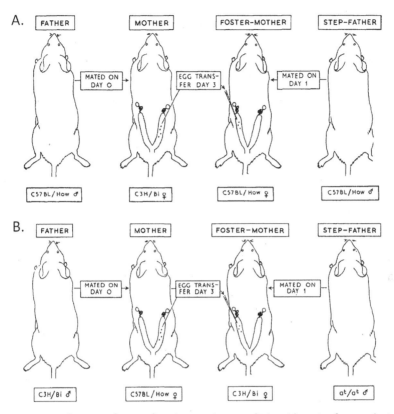

FIGURE 5.4 Structure of a cross-fostering experiment to distinguish uterine from cytological maternal effects. (A) Eggs of donor origin develop into mice with the coat color of the donor strain (C3H/Bi). These mice can then be distinguished from the step-father's offspring that are black C57BL/How mice. (B) A reverse transfer experiment tests the specificity of the maternal effect. The offspring of the a^t male will have a light belly color, allowing them to be distinguished from mice of donor origin. From McLaren and Michie, "Factors Affecting Vertebral Variation in Mice: Experimental Proof of the Uterine Basis of a Maternal Effect."

faithful to the donor's type, or whether they may acquire some of the less desirable qualities of the recipient type."[65]

MATERNAL EFFECTS, APPLIED

From their earliest beginnings in studies of commercial wheat varieties to Hammond's horses, Venge's rabbits, and Fekete's and McLaren's mice, studies of maternal effects were embedded in social and scientific pursuit of technological control over, and optimization of, reproductive out-comes—in agriculture, in the laboratory, and ultimately in the clinic.

The question of maternal intrauterine effects in humans would have to wait. From the 1960s to the 1980s, mammalian maternal effects science found its primary home in the applied genetics laboratories of agricultural scientists seeking to fine-tune the economic productivity of livestock. Agricultural research centers at places like Iowa State University, North Carolina State University, the United States Department of Agriculture (USDA), and Edinburgh and Cambridge in the UK became leading sites for the study of the complex dynamics of maternal effects in heredity. By the 1970s, maternal effects on "parameters of biological and economic interest" were a well-established field of interest in the agricultural sciences.[66]

From a practical perspective, breeders were interested to identify and control maternal effects that might degrade the quality of their breeding stock. They were also keen to improve "maternal productivity." As long as breeders must depend on female uteri to produce new animals, sheep geneticist G. Eric Bradford wrote in 1972, the study of intrauterine maternal effects "will merit substantial emphasis in breeding programs," including the prospect of developing "specialized dam lines with superior performance."[67] In his research, Bradford applied maternal effects science to explore how much ranchers could constrict the size of mature breeding females, thereby "reducing maintenance costs," while still maintaining reproductive quality.[68] A simple economic question guided this line of inquiry: is it possible to breed for maternal "efficiency"—a technical term in the field—that is, to produce females that "mature early, have a high reproductive rate, have very little calving difficulty, stay in the herd a long time, have minimum maintenance requirements and have the ability to convert available energy from forages produced on the ranch into the heaviest possible pounds of weaned calves"?

Beef and swine geneticist Gordon Dickerson, whose career would take him from the Agricultural Research Station in Ames, Iowa, to private industry, and later to the USDA, pioneered the field of applied maternal effects research in the agricultural sciences. In a foundational 1947 study involving 746 pigs tracked over a 12-year period at the Iowa research station, Dickerson examined variation in maternal performance. Dickerson looked at paternal half-siblings (same father, but different mothers) in inbred lines of pigs, measuring offspring efficiency in use of feed to gain weight after weaning. After slaughter, measurements were taken of the composition and dimensions of their total carcass as well as the composition of common consumer cuts: "trimmed ham, trimmed loin and picnic shoulder" and "trimmed belly, fat back and leaf fat."[69]

Dickerson's data showed that pig body composition varied with maternal lineage. In other words, maternal effects were indeed an important constraint in breeding for size and rate of gain in offspring. But there was an unexpected twist. "High-performing" mothers that produced the best rate of gain on low feed requirements in offspring led to dams that were "low-performing" mothers in the next generation. "This suggests," Dickerson wrote, "that the genes which cause pigs of a line to gain more economically," a desirable trait for breeders, "also cause the sows of that line to be poorer mothers."[70] The practical implication of Dickerson's finding was that breeders cannot just select for "rapid and economical fattening," as doing so would soon cause "damage . . . to yield of lean meat and suckling ability."[71] "The net result of this 'see-saw' process," Dickerson concluded, "is exceedingly slow progress over long periods of time."[72] Studies across many species soon seemed to affirm Dickerson's finding that intrauterine maternal effects can have complex antagonistic and inertial intergenerational consequences.[73]

The seesaw of maternal effects across generations presented a challenge for breeders and geneticists. Geneticists portrayed maternal effects as creating "additional complexity" causing "biases in heritability estimates" as well as "inconsistencies" and "very poor prediction" for certain traits.[74] "In even the simplest selection experiment the number of forces which can be operating is not small and the mind boggles at the thought of the ways in which these forces may act," wrote Oscar Kempthorne, the world-renowned British-born Iowa State University geneticist, in a classic 1955 paper on maternal effects. Listing maternal effects as one of nine outstanding "problems to be solved" by the field of quantitative genetics, Kempthorne briefly outlined the daunting challenge of a "logical examination" of the "whole situation" of maternal effects, requiring comprehension of at least "seven parameters."[75] Adorning their papers with swirling arrow diagrams modeling how maternal genetic and environmental effects may alter phenotypic values (fig. 5.5), leading figures

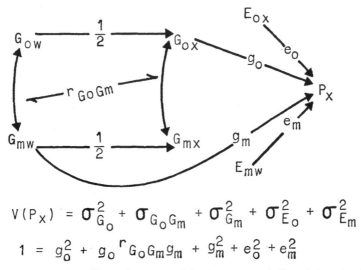

$$V(P_X) = \sigma^2_{G_o} + \sigma_{G_oG_m} + \sigma^2_{G_m} + \sigma^2_{E_o} + \sigma^2_{E_m}$$

$$1 = g^2_o + g_o\,{}^r G_oG_m g_m + g^2_m + e^2_o + e^2_m$$

FIGURE 5.5 A path coefficient diagram modeling how maternal effects (m) might affect phenotypic expression of a trait (P). From Willham, "The Role of Maternal Effects in Animal Breeding."

in mid-twentieth-century quantitative genetics competed to model the varied ways in which a mother's state during gestation alters what would otherwise be predicted on the basis of genetic inheritance.

COULD MATERNAL EFFECTS BE AT WORK IN CONTEMPORARY HUMAN POPULATIONS?

As we have seen, the plausibility of maternal effects in plants, insects, and simple sea creatures gradually came into relief during the first decades of the twentieth century; in nonhuman mammals, it was established in the half-century between the 1930s and the 1970s. Animal studies of maternal effects emerged not from the high-tech laboratories of molecular biology but from the humble world of animal breeding and agricultural research stations. Perhaps in such environments, attachment to theoretical dogma was looser, with a focus more on what actually works for practical application. Through this research, the concept of a "maternal effect" was expanded from the narrow one of early developmental conditioning of seed coats and shell coils to include intrauterine influences in nonhuman mammalian species. By 1967, one scientist described maternal effects as "perhaps the most important source of an indirect genetic effect in mammalian species."[76]

Scientists from Bateson to McLaren recognized that the question of maternal effects goes to the heart of fundamental questions about heredity and reproduction. Against the view that the singular element that matters for development is the DNA that an individual inherits from its mother and father (along with the basic requirements of nutrition and safeguarding from physical harm) they posited that more is transmitted: offspring are shaped as they develop by unknown factors inherent in the maternal constitution that forms the milieu in which they grow. But could such maternal effects be detected in contemporary human populations?

Measuring intrauterine maternal effects is challenging enough in plants and nonhuman animals studied in highly controlled circumstances allowing experimental manipulation and study across multiple generations. Diffuse understandings of just what constitutes a "maternal effect" further complicate the question of maternal effects in humans. The concept of maternal effects captures the intuition that traits are determined by not just genes but also the environment, a component of which is maternal. While Dobzhansky introduced the term "maternal effects" to characterize early embryonic influences of the genetic composition of the mother's cytoplasm in fruit flies, by 1958, McLaren characterized the "term 'maternal effect,'" as referring to any "exceptions to the rule of sex equality, where the route of transmission is non-chromosomal."[77] This extremely broad definition of maternal effects could include everything from cytoplasmic mitochondrial inheritance and maternal influences in early embryo formation, to intrauterine programming of developmental set-points, to postnatal maternal care, socialization, and learning.

Classically, maternal effects refer to more or less permanent constitutional maternal factors. Maternal effects are, as one researcher put it, a kind of "permanent environmental effect" that requires consideration of the "contribution of all the maternal dams back in the pedigree."[78] Some of these factors may be under genetic control. That is, in the mother, there may be "maternal effect genes" that influence the maternally provided environment for offspring, indirectly modifying the phenotypic expression of the offspring's genotype. Other maternal effects may be environmental, produced by the mother's "own nurture, nutrition, and developmental experience."[79] A good example is the mother's own birth weight and eventual adult stature, which will reflect not only genetics but early environmental exposures.

Quite distinct from the permanent or constitutional characters of the mother are more transient, temporary factors, which can vary from pregnancy to pregnancy, such as age, and in humans, smoking, occupation, stress, access to health care, or substance use. Generally, these factors would not be considered maternal effects, but simply environmental exposures to the fetus, conveyed across the placenta. But when such factors are chronic and persistent in the mother—fixed in place, as epidemiologists Jennie Kline, Zena Stein, and Mervyn Susser put it, by "enduring physical environment," social class, "place of residence," and "persisting habits"—perhaps they may rightly be considered within the province of maternal effects.[80]

In the 1970s, anthropologists, sociologists, and maternal-fetal scientists in the fields of nutrition, toxicology, neuroscience, and placentology began to wonder whether social and environmental variables, biologically materialized in offspring via the conduit of maternal effects, could help explain intergenerational patterns of unequal life outcomes in human populations. Their story is the subject of chapter 6.

CHAPTER 6

Race, Birth Weight, and the Biosocial Body

In the 1960s, a particular idea about how to understand the persistence of black American poverty began to trend across a wide political spectrum. Racial disadvantage in the United States, social scientists and political commentators argued, is intransigent because it is transmitted intergenerationally. The implication was that generations of racism, discrimination, poverty, and poor health create entrenched patterns with a course and power that redistributive social policies and civil rights protections can only feebly counteract.

This intergenerational intransigence of racial inequity was attributed by some to culture. A version of this hypothesis animated Senator Daniel Moynihan's 1965 US Department of Labor report, *The Negro Family: The Case for National Action*. The report asserted that African Americans' ongoing disparate social condition originates in a "tangle of pathology" produced by mother-headed households with high rates of out-of-wedlock births. Owing to the legacy of slavery, Moynihan claimed, this cultural pattern of social organization had trapped blacks in a cycle of dependency and inculcated a lack of desire for achievement in black males. Moynihan believed this culturally driven intergenerational cycle explained why so many American blacks remained in the grips of poverty, despite social welfare investments. Acknowledging these cultural patterns resistant to efforts at change, he argued, was essential to devising social policy to promote racial equity.

Others turned to genetics.[1] In a 1969 article in the *Harvard Educational Review*, psychologist Arthur Jensen infamously argued that persistent, intergenerational racial inequities in life outcomes constitutes evidence of differences in genetic propensities related to intelligence. In the 1960s, median IQ scores for American blacks were about half a standard deviation lower than those of whites. Most social scientists believed that the IQ gap between black and white Americans was due to economic and environmental deprivations.[2] Furthermore, they disputed that IQ tests represent valid measures of intelligence across time, social group, and cultural context.[3] But Jensen insisted that the IQ gap represented innate differences in intelligence. Along with other prominent 1960s voices including Nobel Prize–winner William Shockley and psychologist Richard Herrnstein, Jensen argued that the IQ gap was biological, explaining the intransigence of black educational achievement in the face of social investments. Policy makers, he argued, should yield to this reality: social inequities between blacks and whites could not be remedied by more investment in education.

The 1970 book *Disadvantaged Children: Health, Nutrition and School Failure*, authored by Albert Einstein University pediatrician, neuropsychologist, and education expert Herbert Birch and Columbia nutritionist Joan Gussow, stepped into this impasse. The book marshaled a growing body of work by maternal-fetal scientists on intergenerational patterns in low birth weight to launch a critique of Jensen's and Moynihan's pessimistic views of the prospects for racial equality. Intransigent inequalities between groups, Birch and Gussow argued, reflect not genetics or culture but "a continuum of damage" caused by maternal stress and nutritional deprivation, leading to low birth weights and "producing more subtle and widespread damage and resulting in a generally reduced level of mental functioning."[4] The persistent patterns of inequality "are the effects not of the immediate but of the past environment, that in which the mother grew to adulthood."[5] In the debate over the origins of intergenerational patterns of racial inequality, the womb offered an explanatory mechanism for the intractability of unequal life outcomes among American racial groups that blended social and biological explanations.

This chapter examines how, through the convergence of maternal-fetal science and theories of racial progress in the 1960s and 1970s, conceptions of maternal bodies as repositories and vectors for social ills and of fetal bodies as vulnerable and impressionable biosocial beings became a touchstone within larger social debates over racial difference and social inequality. A tour through debates over the causes, remedies, and effects of the birth weight gap between whites and blacks in the United States reveals the complex and intimate entanglement between the sciences of maternal intrauterine effects and the social justice–oriented project of biosocial science in the twentieth century. Still today, the incidence of low birth weight among African American infants is twice that of white American infants, a gap that has remained relatively stable since the earliest records of American birth weights. But researchers no longer believe, as they once did, that the cause is genetic. Social epidemiological science in the mid-twentieth century on the question of race and birth weight forged a different view, of the womb as a space permeated by the wider social environment, and of birth weight as a vivid metric of social inequality and a driver of disparate life outcomes, including educational attainment.

While the locus of concern over low-birth-weight babies began with the aim of reducing infant mortality, interest quickly gathered around the implications of birth weight for debates such as that swirling around genetics and racial differences in IQ. For some, the maternal-fetal interface and its intergenerational implications became a resource for theorizing the harm of racial oppression. The notion that racial trauma could be written on the body and passed through the maternal soma through the generations, trapping black communities in cycles of poverty, operated to move discourse from questions of culture, individual behavior, or innate biology to matters of structural inequality and biosocial inheritance within a social justice narrative of intergenerational transmission of trauma. The vulnerability of even the fetus to the harsh social conditions of blacks in America, and the idea that the social imprinting of racism and poverty begins even before birth, became part of a strategic biopolitics of social justice.[6]

BIRTH WEIGHT

Today, birth weight is a marker of pregnancy success. In the United States, a birth announcement typically includes birth weight, in precise pounds and ounces, along with just a handful of other facts: the name, sex, and length of the child, as well as the date and time of birth. Birth weight statistics regularly appear in national and international data as indices of maternal-fetal health, economic development, and gender equality. In fields ranging from obstetrics to sociology, the variable of birth weight has attracted enduring fascination. A staggering number of scientific publications—more than sixty thousand in just the past three decades, according to a recent literature search—investigate birth weight as a primary variable in a range of health, developmental, and economic outcomes.[7]

Part of the reason for this focus is that socioeconomic factors are strongly associated with low birth weight, and low birth weight is strongly correlated with many poor outcomes in areas ranging from lung, liver, and kidney health to behavior and cognition. It is much less clear whether that association is causal—that is, whether birth weight is itself the cause of those poor outcomes, or whether birth weight is simply a proxy marker along a social trajectory leading to poor outcomes (fig. 6.1). Determining which model is accurate, in combination with the question of what *causes* low birth weight, would point to where the nexus of public health control and intervention might lie. But distinguishing between these two models—birth weight as the pathway to disadvantage, or birth weight as a proxy for social trajectory—has remained a challenge since researchers first began to investigate the social dimensions of birth weight.[8]

Birth weight was not always a biomarker portending such great significance across so many different arenas. Although use of the infant scale began in the eighteenth century in some European maternity hospitals, weight was used to monitor progress in feeding after birth and held little significance beyond that. The first systematic efforts to determine normal human birth weights and growth trajectories began in the early nineteenth century, a practice of quantitative tracking that historian of science Lawrence Weaver argues was linked to the rise of medicalized childrearing under the guidance of a physician.[9] The infant weighing scale arrived rather later in the Anglophone world. Historical medical catalogues suggest that scales began to be routinely marketed to family medical offices in the 1870s, and that birth weight statistics began to be systematically gathered in some hospitals in the United States and Britain around the turn of the twentieth century. However, birth weight did not

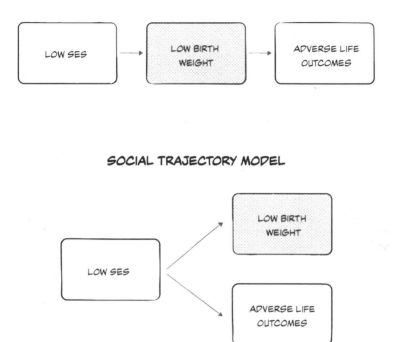

FIGURE 6.1 Birth weight pathway model versus social trajectory model for explaining the relationship between socioeconomic status (SES), birth weight, and life outcomes. Figure by Amy Noseworthy, copyright Sarah Richardson, 2020.

become a routinely reported and standard vital statistic until the mid-twentieth century, when changes in patterns of infant mortality caused it to emerge as an important variable in public health.[10]

An epidemiological shift occurs when a medical innovation, social intervention, or demographic transition leads to stark changes in a population's disease distribution. Although very small babies had always fared poorly, the vast burden of infant mortality before the mid-twentieth century had been the result of infectious disease in the first year of life. Children in late infancy once died of common infectious diseases, such as measles, mumps, rubella, diphtheria, influenza, and wasting diseases such as cholera, at very high rates. But between the 1930s and the 1950s, infectious disease control and vaccines led to a massive decline in these mortality rates in the United States. By 1970 in the US, the rate of child death under age five from infectious disease had declined from 1.7 percent around the turn of the twentieth century to around .2 percent.[11]

Now that deaths in late infancy from infectious disease had become rare, the remaining burden was in the first month and especially the first week of life. In this context, birth weight quickly became salient as a predictor of infant mortality risk. Though infant mortality would remain a matter of great public health interest, epidemiologists and activists began to shift their focus to more fine-grained measures of maternal and infant health. They sought metrics that could predict poor outcomes and that could serve as markers of progress over time. They also needed quantitative measures that pointed to the ongoing need for investment in the field of maternal and infant health. By the 1960s and 1970s, birth weight became a leading-edge biometric of infant health, as well as of national and populational health outcomes.[12]

A second epidemiological shift, from the 1960s to the 1980s, saw a large uptick in live births of increasingly very low-birth-weight infants, due to modern obstetric interventions and new medical technologies, such as incubators, for sustaining life in preterm infants. This increased both the rate of low-birth-weight infants and the apparent rate of infant mortality due to low birth weight. It also amplified the number of low-birth-weight children living into adulthood, spurring medical researchers' interest in the long-term health and social implications of being born too small.[13]

The postwar explosion in research on maternal and fetal determinants of birth weight was facilitated by massive expansions in the data infrastructure for collecting vital statistics data. In the US, the birth certificate was introduced in 1895 as a basic document of citizenship. Over time, it also became a collection point for birth data, developing into a critical tool in vital health statistics and public health. In the 1940s, the standard form began to request medical information as well, mostly about the mother's health at the time of birth. In 1950, a requirement was added to the standard birth certificate to include birth weight. Another revision of the US birth certificate in 1968 requested a range of socioeconomic information about the parent and greater medical information about the newborn infant. These vast data sets were irresistible to public health researchers and social scientists looking to test novel hypotheses about the relationship between birth weight and life outcomes.

Whether collected on birth certificates or by other means, birth weight data showed clear correlations between low birth weight and poor health outcomes for infants. Moreover, it indicated that there were differences in the rate of low-birth-weight births—as well as average birth weights—across racial, ethnic, and national populations. In the US, this stratification tracked color lines: African American babies were far more likely to be designated low birth weight, or smaller than five and a half pounds according to standards issued by the World Health Organization in 1950, than were whites. The finding opened new inquiry into the causes and consequences of the birth weight gap between American whites and blacks.

RACIAL SCIENCE AND BIRTH WEIGHT

Until the 1950s, scientific researchers and physicians widely presumed that different races would have different characteristic measurements, and the leading hypotheses for racial variation in birth weight were biologically based. Racial disparities in birth weight, that is, had long been considered a feature of innate biological difference. Anthropometrics, the study of constitutional differences between races, ethnicities, and nationalities, emerged in the nineteenth century as a central methodology in anthropologists' vision of a systematic and scientific study of humankind. Famously, the practice, the subject of Harvard evolutionary biologist and science historian Stephen Jay Gould's excoriation of racial science, *Mismeasure of Man*, produced elaborate scientific typologies of race, constructed with precise measurements of body build, head shape, and facial angle. Such evidence supported the then-common scholarly and popular view that the peoples of the world could be divided into distinct genetic stock, each with different strengths and weaknesses, and some superior to others. A generation later, these anthropological claims and methodologies would help fuel scientific eugenicists' calls to save the "race" from degeneration through sterilization of the "unfit," restrictions on immigration, and the preferential distribution of public health services to those considered desirable.

Starting in the 1920s, birth weight statistics were collected occasionally, though not systematically, by European and North American anthropologists seeking to document variation among human races.

Anthropometric researchers posited that the children of different races, ethnicities, and nationalities followed distinct growth curves. Physical anthropologists collected birth weight information alongside other standard population data such as height, head and chest circumference, tibia length, and tooth size. Though typically based on small samples of indigent populations, early anthropometric studies of weight at birth consistently claimed that blacks possessed the lightest birth weights of all populations. At the time, scientists did not turn to nutritional deprivation or unequal access to health care to explain this gap. Instead, the consensus was that the birth weight gap reflected a genetic racial difference between blacks and whites. The "norms," they believed, were simply different for black infants.[14]

Medical practitioners carried these data into the clinic. In a 1927 article titled "Weight of Colored Infants," for example, Case Western Reserve University anatomist C. T. J. Dodge advised that "any dispensary caring for a large number of Negro infants should print the colored curve on their graphs, as the growth curves are entirely different from those of the whites." Appropriate recognition of racial differences in birth weight and subsequent growth, he contended, is necessary to prevent physicians from being "misled" into diagnosing malnutrition in small black infants.[15] Similarly, University of Cincinnati pediatrician Robert A. Lyon argued in the 1940s that birth weight differences represent "a true racial characteristic" and advised that physicians "establish separate weight standards for the white and Negro races."[16]

One scientist speculated in 1934 that dark skin was the culprit: "There is a good deal of evidence which would lead one to believe that the pigmented skin of the Negro filters out much of the available ultra-violet light, thus probably playing a part in the production of weakness and lessening birth weight."[17] Another prominent view during this time was that the birth weight gap between American blacks and whites reflected "differences between the gestation periods of the Negro and the white race."[18] Anthropologists claimed that black babies were smaller but hardier than white infants at birth. Indeed, until the 1950s, it was considered common medical knowledge that black fetuses matured faster and that black mothers had a shorter gestation time than whites.

But anthropometrics was also put to other, more progressive, ends. A 1912 study by the American anthropologist Franz Boas, for example, showed that children of immigrants to the United States were markedly taller and had larger heads than their parents. The study, produced for the United States Immigration Commission using anthropometric data from 13,000 immigrants and their children, offered a radical critique of eugenicists' essentialist and hereditarian views of racial difference. Boas' "biosocial" approach to human biological variation suggested that seemingly fixed biological traits can vary depending on social forces, and, optimistically, that each generation presents a fresh opportunity for positive social forces to maximize human potential.[19] This biosocial approach to understanding racial differences in physiological measurements inspired later generations of researchers investigating the origins of the black-white birth weight gap.

THE WOMB AS SOCIAL ENVIRONMENT

In the 1950s, two US research groups located at prominent historically black colleges undertook the first large-scale epidemiological studies of the problem of the black-white birth weight gap. Edward Perry Crump of Meharry Medical College in Nashville, Tennessee, and Roland Scott of the Howard University School of Medicine in Washington, DC, carried out the studies. Crump, an expert in premature birth, was the founding chair of the Department of Pediatrics at Meharry. He became a nationally known expert in the relationship between socioeconomic factors and infant mortality, and to this day Meharry bestows an annual award in his name. Writing on "Negroes and Medicine" in the journal of *Pediatrics* in 1959, Crump bemoaned the "rugged pattern of uphill struggle for initial training opportunities, for hospital staff privileges in the care and follow-up of his patients, and for recognition by and integration into the major medical society (AMA) in this country" faced by black physicians. He could have been writing about Scott, a pioneering figure in the mostly white field of pediatrics who in 1934 became the first African American to pass the pediatric board exam. Scott, who died in 2002, recently made national headlines when, in 2020, the American Academy of Pediatrics at last issued a public apology for rejecting his application for membership and preventing his attendance at society meetings solely because he was black.[20]

Scott and Crump instituted a line of research and debate that continues to this day, constituting birth weight as a prime measure of the effects of poverty and racial discrimination and as a high-priority point of intervention for reparative health justice. Scott, at Howard's Freedmen's Hospital, led the way with a series of studies published in leading pediatric journals.[21] Incredibly, prior to 1950, the largest previous study of average weight at birth by race had included only 187 black births. Scott, by contrast, reported the birth weights of 11,919 black infants delivered at Freedman's between 1939 and 1947. Whereas previous studies in the anthropometric mode had focused on identifying a standard median African American birth weight "norm," presumed to be a biological characteristic of the race, with his comparatively massive data set, Scott was able to highlight significant heterogeneity within this population of births.

Scott analyzed birth weights by income of parents, nutrition, and prenatal care. These analyses clearly demonstrated a positive correlation between income and birth weight: wealthier households had larger babies. On the basis of this evidence, Scott concluded, contra the early anthropologists, that there is no innate biological difference between whites and blacks in birth weight. Scott argued that socioeconomic status, which correlates with access to good nutrition and prenatal care, is the primary factor in higher rates of low birth weight in black infants. Institutional investments in prenatal care and proper nutrition for poor pregnant women, he contended, was necessary to address high rates of low birth weight among African Americans

Working separately at Meharry, Crump published a series of studies carried out at the black-only Hubbard Hospital. Crump aimed his sights directly at the hypothesis that differences in birth weight are genetic. As he wrote, while existing evidence showed that "a quantitative birth weight deficit is characteristic of Negro infants, when compared with whites, . . . papers appearing in the medical literature generally have not reflected a thorough study of the socioeconomic background or level of living of families of these infants."[22] Thus, the finding "that Negro infants are smaller at birth than white babies" cannot be taken as demonstration "that this is exclusively a racial characteristic."

In his studies, Crump reported detailed data on the "socioeconomic index" of the parents of newborns at Hubbard. He focused on the mothers. As he wrote, "All determinants of birth weight act principally through the mother, from conception to delivery, for she provides 100 percent

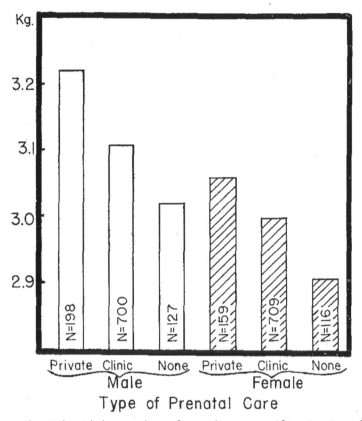

FIGURE 6.2 Birth weight by sex and type of prenatal care among African American infants born at the Hubbard Hospital in Nashville, Tennessee, between 1951 and 1956. From Crump et al., "Relation of Birth Weight in Negro Infants to Sex, Maternal Age, Parity, Prenatal Care, and Socioeconomic Status."

prenatal environmental control."[23] Crump's results showed "a direct relationship between socioeconomic status of the Negro and the birth weight of infants within each socioeconomic group" (fig. 6.2).[24]

Research by Scott, Crump, and others on the social determinants of racial variation in birth weight extended biosocial theories of the determinants of life and health outcomes into the prenatal period. Birth weight took shape as a stark, simple numerical metric representing the reality of the profoundly negative somatic consequences of poverty, oppression, and other forms of deprivation from the earliest stages of development. By 1974, a March of Dimes symposium, "The Infant at Risk," elevated socioeconomic factors as "the prime cause" of low birth weight, issuing a call for research on how "psychologic, cultural, and socioeconomic events" during intrauterine life contribute to "a poor product."[25]

Studies demonstrating the social corollaries of birth weight had a wider intellectual impact, too. From Watson and Crick's deduction of the structure of DNA in 1953, to the first cytogenetic test for Down's Syndrome in 1959 and the revelation of a genetic code for amino acids in 1966, the 1950s and 1960s were a triumphant period for molecular biology. By the 1960s, molecular biologists were advancing a forceful gene-centric vision of biological development as fully determined and governed by DNA.[26] Critics of this gene-centrism pointed to studies of maternal effects as evidence that more is inherited by offspring than just DNA, and that DNA itself can be altered or overridden by other factors. Indeed, the mid-twentieth-century intellectual descendants of Franz Boas frequently invoked maternal-fetal effects as they developed a critique of genetic determinism that conceptualized the body as plastic and biology as contingent on its environmental context.

In his 1956 *The Biosocial Nature of Man*, the anthropologist Ashley Montagu, a student of Boas, placed the intrauterine environment at the center of his case for biosocial science and against genetic determinism. A properly biosocial approach, Montagu wrote, recognizes "three components of human nature," the genetic, the cultural, and—completing the triad—the "uterine environment."[27] For Montagu, intrauterine influences became a resource for expanding the extent of the entanglement of putatively genetic traits with environmental factors. Even if a trait appears to have no environmental influences, one cannot conclude it is genetic, because, Montagu argued, "we should always have to reckon with the history of the individual while he was in the womb."[28] Uterine factors, Montagu claimed in *The Biosocial Nature of Man*, and also in his 1962 *Prenatal Influences* and 1964 *Life Before Birth*, stand as powerful exemplars of how environmental variables are "capable of modifying the expression of the genotype, so that the phenotype . . . presents itself not as an expression of its genetic endowment, but as an expression of the interaction between its genetic endowment and its environment."[29]

The picture of women's bodies as sensitive biological barometers of the social environment given by Crump and Scott suggested that birth weight should be responsive to additional socioeconomic resources, and that higher birth weight should lead to better life outcomes. According to this view, outcomes of interest, like infant mortality or cognitive development, were directly mediated by low birth weight, which could be improved by adding poundage to the growing fetus through superior prenatal nutrition.[30] Epitomizing this view, Columbia nutritionist and neuroscientist Myron Winick claimed in 1972 that "pound for pound, the poor baby does as well as the rich baby; black babies do as well as white babies. The difference in mortality can be entirely explained by the fact that babies from these disadvantaged groups weigh on the average half a pound less at birth than middle-class babies. Theoretically, at least, if we could raise the average birth weight in this portion of the population by a half a pound, we could erase the difference and significantly reduce our overall infant mortality."[31]

But a number of indicators began to suggest that it wasn't that simple, leading researchers to ask whether something else—"maternal effects"— might also be at work.

MATERNAL EFFECTS AND INTERGENERATIONAL INERTIA

The first hint that maternal factors specific to each mother—her health, genetics, and life history, including her own gestational environment— are also at work in integenerational patterns of health and human development came from a survey of infants gestated during the Dutch Hunger Winter famine, published in 1947. The data showed only mild effects of famine on birth weight. Even badly nourished women, it seemed, gave birth to babies with healthy weights.[32] Then, in the 1950s at Vanderbilt in the United States and Aberdeen in the UK, researchers carried out large observational studies of diet during pregnancy in relation to infant outcomes. Surprisingly, both studies failed to show any effect of nutritional intake on birth weight, suggesting that birth weight is not a simple matter of what mothers consume during the pregnancy itself. Despite the "belief which seems generally reasonable" that ample dietary intake during pregnancy produces more robust babies, researchers concluded that there is "no evidence that the diet taken during pregnancy has any special significance."[33] Scientists interpreted these findings as evidence that it is not easy to raise birth weights in a generally well-nourished population, and that factors other than nutrition, perhaps related to the mother's genetic makeup and overall health, built during her own growth and adolescence, are more critical to birth weight outcomes.[34]

Then, a series of studies in the 1960s looked more closely at birth outcomes among educated and upwardly mobile black women in the United States. The results showed that black women in the professional classes, with adequate income and access to good medical care, still had higher rates of low birth weight infants and perinatal mortality than whites, a challenge to Crump and Scott's claims that socioeconomic status of the mother determines low birth weight.[35] Further research revealed that while the rate of low-birth-weight infants had declined among all populations in the immediate period after World War II, starting in the late 1950s the rate stopped improving for African Americans despite heightened social investments in public health and welfare.[36] By 1967, the birth weight differential between blacks and whites was greater than it had been in 1950.[37] This suggested, to some interpreters, that some sort of inertial, intergenerational effect was maintaining lower birth weights in black American populations.

A further challenge to theories that birth weight is a simple barometer of socioeconomic environment arose when scientists began to look more closely at birth weight in relation to gestational age. Prior to 1960, birth weight statistics lumped together babies who were small because they were born prematurely with full-term births that were small—a protocol followed in Crump and Scott's research as well. In the early 1960s, researchers realized that low birth weight caused by prematurity may be epidemiologically and clinically an entirely different phenomenon than that caused by growth restriction in a baby small for its gestational age. In 1961, Johns Hopkins physician and placental pathologist Peter Gruenwald introduced the concept of "intrauterine growth restriction" (IUGR) to describe low birth weight caused by growth inhibition, distinguishing it from low birth weight resulting from prematurity (fig. 6.3). IUGR described the condition of infants who had, for some reason other than obvious congenital defect, failed to reach the expected weight for their gestational age. Whereas a premature baby weighing under 2500 grams may be a typically sized baby simply born too early, most IUGR infants are full-term live births in the lightest decile.[38]

The distinction between preterm and IUGR infants made new patterns clear in the data on inheritance of birth weight, changing

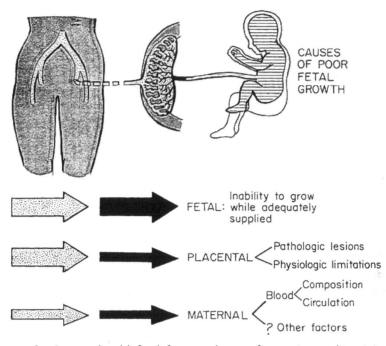

FIGURE 6.3 Conceptual model of a role for maternal causes of intrauterine growth restriction. From Gruenwald, "Fetal Deprivation and Placental Insufficiency."

understandings of the causal mechanisms underlying the relationship between maternal factors, birth weight, and poor life outcomes. Socioeconomic factors such as those identified by Crump and Scott well explained low birth weights caused by preterm birth. But, when looking only at IUGR births, with preterm births removed from the picture, "maternal effects" appeared to play a large role in birth weight outcomes. Up to 50 percent of the variation in IUGR birth weight, it seemed, was explained by the mother's—but not the father's—constitutional qualities.[39] For example, mothers who were themselves small at birth had smaller babies. Some intrauterine environments, it seemed, were inherently less nourishing or more debilitating than others—or, as agricultural scientists studying the same phenomenon in livestock had put it—they were "low-performing," less "productive," or less "efficient."

As a consequence of these findings, in the 1960s and 1970s, biosocial theories of birth weight variation began to shift. While researchers continued to recognize that social factors could directly affect birth weight particularly in the case of preterm births, low birth weight in the full-term infant presented a more complex picture of the maternal body as an inertial constraint and mediator of social influences, through cryptic processes not yet physiologically understood. The authors of the Aberdeen study of diet in pregnancy hypothesized that the lack of evidence for gross effects of nutritional deprivation or supplementation on birth weight suggests that researchers must look to "the idea of impaired [maternal] efficiency rather than of breakdown."[40] Perhaps, they proposed, "the key will . . . be found in the intimate processes of [the mother's] physiological adaptation to pregnancy."[41]

Leading the way in this conceptual shift toward embracing a central role for maternal effects in birth weight and associated outcomes were doctors Margaret and Kit Ounsted of the Park Hospital for Children in Oxford, UK. Pediatricians and mouse researchers who were experts in intrauterine growth, the husband-and-wife team influentially hypothesized that a mother who herself experienced intrauterine growth restriction would pass this birth weight effect on to her own offspring, creating an intergenerational effect of "growth constraint" in the maternal line.[42] "It is apparent," they wrote in their book, *On Fetal Growth Rate*, "that in some women there is a constraining mechanism which is pre-potent, and regularly controls the growth rate of all their young." "We [suspect] that the limits of constraint might in part be determined by the degree of constraint imposed on the mother when she herself was a fetus." The theory predicted that environmental lesions during intrauterine life, such as those originating in socioeconomic deprivation as experienced by black women in the United States, might grip a family for generations after the exposure had passed.[43]

Ounsted and Ounsted hypothesized that such maternal effects function as a non-genetic form of adaptation permitting plasticity in response to rapidly changing conditions. Through maternal effects, they argued, "The matrilineal line has many openings for the transmission of information down the pedigree, in addition to those coded in the genome." With respect to widely observed differences between racial groups, "maternal constraint," they proposed, "operates within ethnic groups to draw mean growth rate towards that which is most adaptive for the particular gene pool as it varies over historical time."[44]

Citing the Ounsteds, Johns Hopkins placenta expert Peter Gruenwald

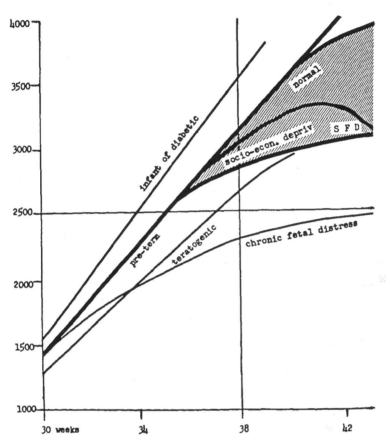

FIGURE 6.4 Model of the contribution of socioeconomic deprivation to intrauterine growth restriction. The y axis represents birth weight in grams. From Gruenwald, "Fetal Deprivation and Placental Pathology: Concepts and Relationships."

hypothesized that the study of "intrauterine life" "can give an indication of the overall extent of [socioeconomic] deprivation and may be a sensitive indicator of change" (fig. 6.4).[45] Gruenwald further theorized that the "delayed" effects of what he termed "fetal deprivation" meant that social interventions should be directed not at pregnant mothers, per se, but with a longer-term, intergenerational horizon in mind. "This strongly suggests that the full manifestation of socioeconomic improvement may take two or three generations," he opined in the journal *Obstetrics and Gynecology*.[46]

Ounsted and Ounsted's maternal constraint theory, and speculations like those of Gruenwald, might have remained an intriguing nugget for the consideration of a small audience interested in the interaction between heredity and pregnancy. But just as researchers across diverse fields were beginning to theorize how maternal constraints might produce intergenerational inertia in birth weight, questions about limits on the pace of change and persistent intergenerational patterns in racially stratified human populations were being raised in the realm of debates over genetic and social factors in racial differences in IQ and educational attainment by figures such as Moynihan and Jensen. It was not long before some researchers proposed that maternal effects might help explain these patterns.

MATERNAL EFFECTS AND BIOSOCIAL THEORIES OF RACIAL GAPS IN LIFE OUTCOMES

While interest in birth weight began as a way to tackle higher rates of infant mortality among populations with high rates of underweight infants, in the 1960s and 1970s nutritionists began to expand their inquiry to examine the broader consequences of birth weight as a continuous variable. That is, they extended claims about the negative effects of low-birthweight infants, defined as those below five and a half pounds, to variation within the "normal" spectrum of birth weight, including six and seven pound infants. Even a below-average birth weight, they hypothesized, may produce significant, as-yet-undetected health and cognitive outcomes.

Prenatal researchers across the nutrition, brain, psychology, and education fields connected clams about the intransigence of achievement gaps between racial groups to deprivations during the early development of the brain in the womb. Harvard prenatal neuroscientist Abraham Towbin, for instance, speculated in 1969 that "subclinical minimal cerebral damages are of common occurrence, and in such cases 'the potential of performance may be reduced from that of a genius to that of a plain child, or less.'"[47] At a 1972 Society for Neuroscience meeting, Winick argued that although "poverty-stricken" women in the United States "do not look malnourished," their "chronic borderline malnutrition" results in lower birth weights, harming children's "brain development."[48]

UCLA neuroscientists Edith van Marthens and Stephen Zamenhof, drawing on their experiments on the effects of malnutrition on rat brains, explicitly linked maternal intrauterine effects to the intergenerational intransigence of efforts to close the IQ gap between black and white Americans. Mild undernutrition and reduced birth weight in humans, they hypothesized in the journal *Molecular and Cellular Biochemistry*, may underlie the "more numerous, and therefore more important for society . . . borderline (subclinical) cases" of low IQ individuals.[49] "The implications of these findings must be carefully considered in studies of genetic factors and intelligence," they wrote. "For example, animals who are continuously malnourished [during gestation] show progressively greater learning deficits across successive generations." Analogizing this to the human case, van Marthens and Zamenhof suggested that "the beneficial results of feeding programs for the offspring of malnourished members of our society may become progressively apparent only after several generations have passed."[50] Drawing on maternal constraint theories of the sort first proposed by Gordon Dickerson in 1947 for livestock, and expanded by Ounsted and Ounsted and Gruenwald in the 1960s and 1970s for humans, they argued that "non-Mendelian, maternal inheritance" produces "long-range regulatory mechanisms" that "transfer . . . the effects of prenatal malnutrition in one generation, on brain development in the next generation." The deficits need not be large; even "cryptic malnutrition may result in inheritable learning deficits . . . which may be non genetically transmitted to the next generation."[51]

Birch and Gussow's *Disadvantaged Children: Health, Nutrition and School Failure* synthesized these maternal-fetal theories of the origins of inequality for social scientists, policy makers, and the general public. Achievement gaps between racial groups in the United States, Birch and Gussow argued, are enacted at the material level of the body, but they are social in origin. To "break the continuous intergenerational chain of poverty," the authors proclaimed, summoning Montagu's biosocial doctrine, "society can no longer afford to deal with cultural influences as if they were acting upon a nonmaterial being."[52] As long as opportunity is not equalized in the womb, Birch and Gussow warned, "conditions of life, both past and present" will "make certain groups of mothers less efficient child-bearers," and "intervention at a single point must inevitably have a limited effect."[53] With the theory of maternal constraint of birth weight at the center of their argument, Birch and Gussow argued that to improve educational outcomes, society must make biological as well as social interventions: "there is no reason to think we can fully compensate the child handicapped by an existing biologic disadvantage merely by increasing his objective opportunities for learning in school settings."[54]

Early biosocial theories had posited the womb as itself a social environment. On this account, the fetal body is directly exposed to the same pervasive present-day social and environmental inequities experienced by the mother, and the highly plastic and sensitive metric of fetal growth, measured as birth weight, represents a barometer of social welfare. According to these theories, the right course of action is to alter structural inequalities experienced in the womb, principally by delivering greater resources to pregnant women and reducing environmental toxins in human communities. But as the complexity of birth weight patterns in human populations came into full relief, a different sort of biosocial theory arose. Bringing maternal effects theories forged by the mouse researchers and agricultural breeders discussed in chapter 5 into theories of human birth weight dynamics, scientists reconceptualized women's reproductive bodies not merely as conduits for social and environmental exposures, but as repositories of social markers and as amplifying vectors for social ills, creating inertia in social change.

There are deep tensions in biosocial theories that relate maternal-fetal factors in birth weight variation to racial inequalities in areas such as educational attainment. Rather than offering the promise of improvements through environmental enrichment, these theories suggest that social interventions targeted at pregnant women might be fruitless: some wombs are destined to produce more vulnerable and less socially optimal offspring. In this way, theories of the fetal origins of unequal outcomes can become another form of evidence for the intransigence of racial inequalities and their ultimate locus in biology. All that can be done in that case, lamented theorists, is to tailor interventions to the biologically lesioned individuals. Thus, even as progressive biosocial theorists such as Birth and Gussow resisted genetic determinist claims like Jensen's, they arrived at other forms of determinism rooted in the conditions of the womb.

THE PUZZLE OF BIRTH WEIGHT

By the end of the 1970s, scientists were churning out more than two thousand papers a year on low-birth-weight babies.[55] Four presuppositions grounded the biosocial theory of birth weight variation and racial stratification in life outcomes that emerged in the 1960s and 1970s. First, that prenatal development represents a critical period of high plasticity in which environmental variables can produce large effects. Second, that birth weight is a reliable signal of maternal provisioning in the womb, which may be influenced by both the proximate and accumulated social-structural well-being of the mother. Third, that low birth weight is itself a severe biological lesion that only vigorous social interventions can hope to correct, potentially over the course of generations. Fourth, that knowledge about the health risks of being low birth weight carries insights for research on potential risks linked to variation in birth weights in the normal range, or between mean birth weights among populations. Each of these assumptions underwent serious scrutiny and reevaluation during the 1980s and 1990s.

The denouement of the hypothesis that prenatal undernutrition drives group differences in areas like IQ and educational attainment began with the 1975 publication of the results of the first follow-up study of gestational survivors of the Dutch Hunger Winter famine. The Dutch Hunger Winter was an acute four-month famine at the end of World War II among an otherwise healthy population in the Netherlands. The famine resulted in 10,000 deaths. Careful nutritional records trace with great precision the degree of malnutrition over the course of the famine. At times, people lived on fewer than 400 calories per day. On average, individuals in the affected population lost 15 to 20 percent of their body weight. Birth records suggest that some 40,000 individuals experienced the famine in utero.[56]

It is little known that this widely cited study of the long-term effects of prenatal undernutrition among survivors of the Dutch Hunger Winter was initiated in order to investigate hypotheses about the role of prenatal nutrition in adult IQ. In 1967, Zena Stein and Mervyn Susser, nutrition researchers at Columbia University, who were keenly interested in the question of intransigent inequalities in educational attainment between blacks and whites in the United States, recognized that gestational survivors of the Dutch Hunger Winter offered a unique opportunity to examine the relationship between prenatal undernutrition and adult IQ. To epidemiologically investigate "the environmental causes of the polarities in the social distribution of mild degrees of mental retardation,"[57] Stein and Susser scoured hospital records and population registers across the Netherlands. They identified a cohort of 2,000 22-year-old Dutch men exposed in utero to the Dutch famine, for whom both birth weight records and military induction records including IQ and mental health measures at age 18 were available. The researchers hypothesized that critical periods of brain growth occur during gestation, that nutritional deprivation affects this brain growth, and that therefore survivors of the Dutch Hunger Winter would have depressed IQ scores. With this study, the Columbia researchers set out to challenge Shockley, Jenkins, and Rushton, who had posited that "disparities in mental performance among social groups" are due to genetics, and to show instead that environmental factors such as prenatal nutrition may underlie these social patterns.[58]

Today studies of the cohort of infants gestated during the Dutch Hunter Winter have expanded into multiple lines of research, standing as a departure point for a vast field of research into the developmental origins of health (the subject of the next chapter). Given this, it is often forgotten that the first Dutch Hunger Winter study failed to find any predicted effects at all. In their 300-page 1975 book *Famine and Human Development: The Dutch Hunger Winter of 1944–1945*, now a classic in the field of epidemiology in part for its extended and self-critical report of negative results, the authors reported that "no association could be demonstrated between prenatal nutrition and all the measures of mental competence available to us."[59] Despite access to "extensive, reliable, and valid data" on prenatal exposures with a degree of "precision unknown for any other sizable human population," no correlation between prenatal nutrition, birth weight, and "the social distribution of mental competence" could be detected among gestational survivors of the Dutch Hunger Winter.[60]

Many other studies would attempt to link birth weight and IQ, but there were endemic problems with such research. A 1989 *Journal of Pediatrics* metanalysis of 80 studies of low birth weight and IQ found that the research was full of small effect sizes of little actionable significance. Mean IQ differences between low birth weight and control children were only three points lower, meaning that as a group, infants born small did not have lower IQ scores of the sort that would explain life outcome disparities.[61] The metanalysis also found that studies did not give enough time, when assessing outcomes, to allow evaluation of whether the purported effects were lasting or attenuate as children develop (there is ample evidence that small size at birth itself, under conditions of adequate nutrition, is quickly overcome by catch-up growth leading the baby to reach its genetically endowed size). The metanalysis concluded that "multiple issues in the burgeoning follow-up literature" on birth weight and IQ, educational attainment, and other life outcomes so compromise the science that they "preclude drawing firm conclusions" about the "ultimate outcome" of being born small.[62]

"The emphasis on birth weight," Stein and Susser would write in their 1989 book *Conception to Birth: Epidemiology of Prenatal Development,* "could be said to have been exaggerated."[63] Far more informative for outcomes ranging from infant mortality to educational attainment than birth weight alone is the gestational age of the baby in relation to birth weight, along with certain information about the ecological backdrop of the infant's fetal growth. One key piece of evidence was the observation that being low birth weight has a stronger connection to infant mortality in more developed countries than in developing countries. In countries such as the US, low-birth-weight babies at term were rare. Those born small were generally preterm births. In the developing world, there were much higher rates of low-birth-weight babies, but they tended to be full term, and when they were born live, a larger proportion of them survived. When babies were small-for-term in developed countries, this was usually due to late pregnancy effects, and the result was a skinny baby with a normal skeletal structure. In developing countries such as Guatemala, growth restriction seemed to begin in the first trimester and led to symmetrical effects: stunting, small skeletons, and thin babies, outcomes that were attributed to chronic malnutrition during the first trimester.[64] In short, how a newborn baby got to be small mattered.

Meanwhile, negative results in experimental studies designed to raise birth weights through supplementation or other interventions piled up. For seven years, from 1969 to 1976, the same Columbia University researchers who pioneered the Dutch Hunger Winter studies tracked black pregnant women in New York City who were randomly assigned to three groups, two receiving supplementary beverages—one high protein and one high calorie—and one receiving none at all. While women taking the supplements gained more weight, this did not translate into increased birth weight. The study found no statistically significant effects on birth outcomes in the supplemented groups, and even found a negative effect in the form of higher infant death rates and lower birth weights in the arm of the study using a high protein supplement. "It is clear to us that a number of our preconceived ideas about maternal nutrition and fetal growth must be abandoned or at least modified," concluded the authors.[65] Another study, which ran from 1980 to 1985, attempted to raise birth weight by providing women with information and counseling about nutrition prior to conception. Alarmingly, the study found that those who received the intervention had more adverse outcomes, including lower birth weights, than those who received no special counseling at all.[66]

While a 1970s supplementation study in a malnourished population in Guatemala had shown the possibility of improving birth weights with extra calories, on balance, supplementation studies proved that the Vanderbilt, Aberdeen, and Dutch Hunger Winter studies had been right: birth weight is not strongly related to prenatal nutrition.[67] As a later commentator would put it, "The relationship between maternal nutrition and fetal growth is nonlinear; deprivation must pass a threshold level before birth weight is significantly affected."[68] Researchers concluded that in well-nourished populations, the contributors to variation in birth weight are difficult to ascertain, and in any case unlikely to be easy to manipulate without invoking unknown risks.[69]

In the 1990s, researchers also made progress on the multidimensionality of the question of the birth weight gap between whites and African Americans in the United States. The gap in birth weight between whites and blacks has been noted for more than a century, as evidenced by nineteenth-century hospital birth records in Boston Lying-In Hospital and Johns Hopkins Hospital in Baltimore showing a 200-gram difference between the average birth weights of blacks and whites, with double the rate of low-birth-weight infants in blacks compared to whites—statistics similar to those seen today. Researchers discovered, however, that this birth weight gap was driven by different factors in different periods. In the early twentieth century, higher rates of premature births, resulting from much higher rates of infectious diseases among black women than women of other social groups at that time, explain much of the difference between black and white average birth weights. In contrast, by 1988, being "unmarried" had become the strongest predictor of having a low-birth-weight baby among black women. This, of course, is a difficult-to-interpret marker that itself may capture a broad range of life history and socioeconomic variables, dependent on social and cultural context—and a demographic trend only salient since the 1960s.[70]

The epidemiology of the racial gap in birth weight in the United States has shifted over time, including in recent decades. An examination of changes in birth weights between 1950 and 1990 in the state of Illinois found that average birth weights increased among both whites and blacks over this time period, but that white birth weights grew at a faster pace. Whereas in 1950, white babies were on average 230 grams heavier than black babies, in 1990, they were 256 grams heavier. One major factor in this pattern was a 56 percent increase in "very low-birth-weight" infants (tiny babies under 1500 grams, or about 3.3 pounds). Today, many of these babies survive who would have been stillborn in the past. By comparison, such births became rarer in white populations during this same period.[71]

Further study has affirmed early findings suggesting that class does not completely explain the black-white birth weight gap. College educated black women in the United States still have an elevated risk of having a low-birth-weight baby compared to their white peers. Researchers believe that experiences of racism as well as culturally specific variables explain the persistence of high rates of low-birth-weight infants in African American women, regardless of their class. For example, a 2006 study showed that among poor black women the presence of a grandmother living in the home reduced the risk of having a low-birth-weight baby by 56 percent, a correlation not found among white women.[72] The researchers hypothesized that as black women become upwardly mobile, they lose these extended kin networks. At the same time, high socioeconomic status black women more often live as minorities in whiter worlds, encountering a higher intensity of negative affective responses to their blackness and more frequently confronting economic and institutional racism. The stress of coping with this may negatively impact their health status and that of their children.[73] The role of extended kin networks in mediating birth weight, combined with the complicated relationship between class and the health impacts of encounters with racism, underscores the complexity and profound ecological dependence of the variable of birth weight.

This complexity is further highlighted by the case of patterns in birth weights among first-generation and second-generation immigrants in the United States. Excluding Puerto Ricans, since 1960 the weight of newborn infants of first-generation Hispanic American immigrants has been more similar to white Americans than to blacks, despite the fact that Hispanic immigrants share a similar high-risk profile to US blacks in terms of health exposures and experiences of discrimination and poverty. By the second generation, immigrants have low-birth-weight rates more similar to non-Hispanic blacks in the US. Clearly, something is different about the embodied experiences of Hispanic immigrants compared to other populations that is protective against poor birth outcomes, and something happens between the first and second generation of immigrant families to alter this protective effect. The experience of Hispanic immigrants to the United States makes clear that factors specific to a given population, and to the life histories of individuals in that population, are critical to understanding the etiology of low birth weight.[74]

The question of whether low birth weight is merely a biomarker of another cause associated with infant mortality and other poor outcomes, or is itself a cause of adverse outcomes, independent of the effects of socioeconomic status, remains a matter of debate today.[75] Undoubtedly, the answer is complex. A 2001 review focusing on the question of birth weight patterns among Hispanic immigrants in the United States found a "bewildering variety of research designs. . . . with contrasting outcome measures, definitions, samples, target populations, methods, and levels of sophistication."[76] In a 2005 review of the developmental origins theory, Harvard pediatrician and nutritionist Matthew Gillman concluded that "birthweight is easily measured and is available from historical records, but if the truth be told, is a dreadful marker of prenatal etiologic pathways."[77] Princeton economist Janet Currie is among those who maintain that being born low birth weight has important and measurable effects on long-term "self-reported health status, educational attainment, and labor market outcomes."[78] In many cases, however, these effects can be erased by being born in a higher socioeconomic class.[79]

In a 2001 essay in the *International Journal of Epidemiology* titled "On the Importance—and the Unimportance—of Birthweight," public health researcher Allen Wilcox argued that after decades of research on birth weight, there is no evidence for a causal link between birth weight and health indicators.[80] "The category of 'low birth weight' . . . is uninformative and seldom justified," wrote Wilcox.[81] For instance, as infant mortality has fallen, mean birth weights have stayed the same. Furthermore, low-birth-weight babies in high-risk populations have lower mortality than low-birth-weight infants in better conditions.[82] Similarly, smokers' babies, twins, black infants, and infants born at high altitudes are all more likely to be born small, but these babies do not seem to have the mortality risks generally associated with being low birth weight. Researchers would do better, Wilcox argued, to turn their attention to preterm birth, a far more telling metric, and a more powerful driver of poor outcomes, including infant mortality, than birth weight.[83]

But many continue to view birth weight as an imperfect but indispensable indicator for the effects of otherwise difficult-to-specify biosocial processes, including the impact of racism on human bodies. In response to Wilcox, physician Richard David has argued that variation in birth weight distributions across different ethnic groups shows that even traits evident from a very early age could already be influenced by environmental deprivations, an observation particularly important "in the context of a society whose dominant elements justify their positions by arguing the genetic inferiority of those they dominate."[84] Birth weight, David argued, is a simple number that may track, better than any known measure, aspects of racism's somatic impact on life outcomes. "Birth weight is collected accurately for almost every in-hospital birth," offering a comparison metric across classes, despite level of access to health care.[85] If we let go of birth weight and the strong patterns of disparity it suggests, "real social problems may be defined away rather than vigorously opposed," David argued.[86] If birth weight plausibly offers a summative physiological measure of biosocial processes during intrauterine development, we are ill advised to relinquish it as a primary measurement, even if hard-nosed analyses find its true meaning elusive. "In applying an analytical technique to real people, to people living in real societies, to people experiencing births and deaths in their families, the scientist must always proceed with an eye on the social, historical and political context in which those real people—and the scientist—find themselves," wrote David.[87]

The story of how birth weight research came to position maternal intrauterine effects as an enriched site for the biosocial inscription of racism on the body proves especially illuminating as we now turn our attention to the rise of intensive interest in maternal-fetal effects today, in the twenty-first-century postgenomic sciences. In the 1960s and 1970s, scientists informed by biosocial theory strategically deployed the science of maternal intergenerational effects on birth weight within larger social debates in the United States over race, social inequality, and biology. The birth weight story shows how, historically, the maternal-fetal interface has long been conceptualized as uniquely revealing the embodiment of social inequalities and the inadequacies of hereditarian thought.

The case of birth weight also carries lessons about the limitations of biosocial research situated at the maternal-fetal interface. First, it reveals how hard it is to study maternal effects in humans. Birth weight proved to be a crude and indirect measure of gestational experience, and research on birth weight failed to yield biological mechanisms linking exposures, birth weight, and later outcomes. The effects, if any, of birth weight variation on later health outcomes are cryptic: they are small, vary depending on social and developmental context, and are embedded within temporally long, multiply confounded, complex chains of causality that make specifying the role of birth weight itself challenging.

Furthermore, it shows how even as biosocial researchers profess a commitment to equality and social justice, powerful raced and gendered ideas—about risk to fetuses and maternal responsibility, and about optimal life outcomes and social futures—can become embedded in biosocial conceptions of the role of bodies and biologies in sustaining social inequalities.[88] As philosopher Jan Baedke and anthropologist Abigail Nieves Delgado have argued, biosocial theories of health disparities in relation to environment and social status characterize racialized populations as "unhealthy, 'at-risk,' and biologically deprived," producing "epigenetic racial distinctions" that "work along parameters such as normality-abnormality, health-disease, modern-primitive, and advantaged-deprived."[89] Moreover, in maternal effects theories of birth weight variation across populations, the mother often represents a constraining, inertial environment invested with causal agency in a vast range of undesirable individual and social outcomes. I explore these tensions in the remaining chapters of this book, which closely and critically examine the science of the maternal imprint in the twenty-first century, in the fields of Developmental Origins of Health and Disease and maternal-fetal epigenetic programming.

CHAPTER 7

Fetal Programming

In 2008, a team of researchers led by Bastiaan Heijmans at Leiden University in the Netherlands and L. H. Lumey at Columbia University in New York City applied a new assay to blood samples from people gestated during the Dutch Famine of 1944–1945. They tested the survivors' DNA for levels of methylation, a molecular mechanism that interacts with DNA to regulate gene expression. Such mechanisms are the focus of the relatively new field of epigenetics, which explores the question of how DNA is turned on and off so as to create diverse forms of life, specialized tissues, and states of disease and health.[1] Almost six decades after the Dutch Famine, Heijmans and Lumey found an average of 5.2 percent lower methylation levels in a gene involved in growth and metabolism among individuals exposed to the famine around the time of conception, compared to their unexposed siblings.

"These data," the authors wrote, "are the first to contribute empirical support for the hypothesis that early-life environmental conditions can cause epigenetic changes in humans that persist throughout life."[2] They argued that epigenetics now offers scientists a measure of "compromised prenatal development" far superior to the "poor surrogate" of low birth weight.[3] The study, which as of 2018 had been cited 1,366 times, became a touchstone in the field of fetal programming research.[4]

The science of maternal-fetal epigenetic programming is a thriving area of human epigenetics research, emerging over the past decade, that investigates how exposures during the prenatal and perinatal periods can induce long-lasting epigenetic changes that lead to adult disease and could be passed on to future generations. Fetal programming researchers believe that perturbations in the fetal environment alter methylation levels, permanently programming gene function in ways that lead to adverse outcomes in children and adults. This causal hypothesis explicitly framed Heijmans and Lumey's study of the epigenetic traces of prenatal famine experienced during the Dutch Hunger Winter. "Epigenetic mechanisms," they wrote in the paper's opening paragraphs, ". . . underlie the relationship between adverse intrauterine conditions and adult metabolic health."[5]

The central working hypothesis of investigations at the intersection of fetal origins and epigenetics is that previously underappreciated variations in the maternal-fetal environment may have long-term, albeit subclinical, effects that carry significance at the population level for the advancement of public health. Previous eras of research on birth anomalies and low-birth-weight infants focused on severe, gross prenatal deprivations causing visible, macroscopic outcomes such as congenital conditions or very low birth weight. By contrast, fetal programming research, using methylation variation as a central tool, poses the specter of invisible, subclinical damage in the form of small perturbations at the maternal-fetal interface with potentially long-term and even intergenerational implications.

Tracking the introduction of epigenetic approaches into maternal effects science, this chapter and the next analyze causal claims in three touchstone research streams that represent some of the most prominent studies linking epigenetic markers with intrauterine exposures and later-life health in human populations. These research streams are summarized in table 7.1: studies conducted from 2008 to 2018 by Heijmans and Lumey, on individuals gestated during the Dutch Famine; research led by Suzanne King between 1998 and 2016, which examined individuals prenatally exposed to a 1998 ice storm in Quebec; and Rachel Yehuda's 2016 study of the offspring of Jewish Holocaust survivors, first mentioned in the introduction to this book. All three research programs emerged from established studies that presaged the introduction of epigenetic technologies and theories. Upon the arrival of epigenetic approaches,

TABLE 7.1. Studies of Maternal-Fetal Epigenetic Effects in Human Populations Discussed in This Chapter

STUDY STREAMS	CENTRAL FINDINGS	NUMBER OF STUDY SUBJECTS	TIMING OF EXPOSURE	DEFINITION OF EXPOSURE	OUTCOME MEASURE	TISSUE	CANDIDATE GENE OR WHOLE GENOME?	EXCLUDE REVERSE CAUSATION?	DATA ON FATHERS?
Dutch Hunger Winter Families Study									
Heijmans et al. (2008) "Persistent Epigenetic Differences Associated with Prenatal Exposure to Famine in Humans," *Proceedings of the National Academy of Sciences*	Periconceptional famine exposure was associated with 5.2% lower methylation in the IGF2 gene compared to controls. Late exposure to the famine was not associated with any methylation differences.	60 individuals as conceived during the famine, 62 exposed only during late gestation, and 122 sibling controls	Periconception	Conception during the famine	Differences in methylation in the differentially methylated region of the IGF2 gene	Peripheral whole blood cells	Candidate gene	No	No
Tobi et al. (2015) "Early Gestation as the Critical Time-Window for Changes in the Prenatal Environment to Affect the Adult Human Blood Methylome," *International Journal of Epidemiology*	Famine during gestational weeks 1–10 was associated with DNA methylation alterations at 4 CpGs. The four CpGs showed a 2.3% increase, 2.3% decrease, 0.7% increase, and 2.7% increase in expression. Two other sites showed methylation changes associated with exposure any time during gestation. One site was specific for exposure around conception.	348 famine-exposed individuals, and 463 time/sibling controls	Periconception or gestation during one of four 10-week periods defined by the researchers	Gestation during the famine for at least one 10-week period or conceived during the famine	Differential methylation at CpG sites on the Illumina array	Peripheral whole blood cells	Whole genome	No	No

(continued)

TABLE 7.1. (continued)

STUDY STREAMS	CENTRAL FINDINGS	NUMBER OF STUDY SUBJECTS	TIMING OF EXPOSURE	DEFINITION OF EXPOSURE	OUTCOME MEASURE	TISSUE	CANDIDATE GENE OR WHOLE GENOME?	EXCLUDE REVERSE CAUSATION?	DATA ON FATHERS?
Tobi et al. (2018) "DNA Methylation as a Mediator of the Association between Prenatal Adversity and Risk Factors for Metabolic Disease in Adulthood," *Science Advances*	Famine survivors' higher BMI and triglyceride (TG), but not fasting glucose levels, are mediated by epigenetic mechanisms. One CpG mediates 13.4% of the association between prenatal famine exposure and BMI and six CpGs together mediate 80% of the association between prenatal famine and TG.	348 famine-exposed individuals, and 463 time/sibling controls	Gestation	Gestation during the famine	BMI, cholesterol, LDL/ cholesterol, from medical examination, glucose from blood draw	Peripheral whole blood cells	Mixed: mediation analysis of candidate CpGs sourced from whole genome study	No	No
Project Ice Storm									
Cao-Lei et al. (2014). "DNA Methylation Signatures Triggered by Prenatal Maternal Stress Exposure to a Natural Disaster: Project Ice Storm," *PLOS One*	Mother's objective hardship correlated with differential methylation in 1675 CpGs in 957 genes related to immune function.	34 adolescents; no controls	Gestated during or conceived within 3 months of the storm	Degree of objective stress reported by the mother on a 32-point questionnaire of objective hardship during the storm, taken 6 months after the storm	Differential methylation at CpG sites on the Illumina array	CD3+ T cells isolated from peripheral whole blood samples	Whole genome	No	No

Cao-Lei et al. (2015) "Pregnant Women's Cognitive Appraisal of a Natural Disaster Affects DNA Methylation in Their Children 13 Years Later: Project Ice Storm," *Translational Psychiatry*	2872 CpGs associated with 1564 genes and 408 biological pathways differed between children of mothers who had a positive vs. negative cognitive appraisal of the ice storm. 793 of these genes overlapped with those identified in 2014, but 771 were unique to cognitive appraisal.	31 adolescents; no controls	Gestated during or conceived within 3 months of the storm	Mother's binary response to one question on a 22-point questionnaire of subjective hardship during the storm, taken 6 months after the storm, asking whether overall consequences of the storm on them and their families were positive or negative	Differential methylation at CpG sites on the Illumina array	CD3+ T cells isolated from peripheral whole blood samples	Whole genome	No	No

(*continued*)

TABLE 7.1. (*continued*)

STUDY STREAMS	CENTRAL FINDINGS	NUMBER OF STUDY SUBJECTS	TIMING OF EXPOSURE	DEFINITION OF EXPOSURE	OUTCOME MEASURE	TISSUE	CANDIDATE GENE OR WHOLE GENOME?	EXCLUDE REVERSE CAUSATION?	DATA ON FATHERS?
Cao-Lei et al. (2015) "DNA Methylation Mediates the Impact of Exposure to Prenatal Maternal Stress on BMI and Central Adiposity in Children at Age 13½ Years: Project Ice Storm," *Epigenetics*	Higher methylation of CpGs reduced the adverse effect of objective prenatal maternal stress (including life events not related to the storm) on children's waist-to-height-ratio and BMI.	31 adolescents; no controls	Gestated during or conceived within 3 months of the storm	Degree of objective stress reported by the mother on a 32-point questionnaire of objective hardship during the storm, taken 6 months after the storm	BMI and waist-to-height ratio at age 13.5	CD3+ T cells isolated from peripheral whole blood samples	Mixed: mediation analysis of candidate CpGs sourced from whole genome study	No	No
Cao-Lei et al. (2016) "DNA Methylation Mediates the Effect of Exposure to Prenatal Maternal Stress on Cytokine Production in Children at Age 13½ Years: Project Ice Storm," *Clinical Epigenetics*	6 CpGs, located in 6 different genes, reduced the adverse effect of objective prenatal maternal stress on cytokine production. 3 of these CpGs were increased in methylation, 3 decreased in methylation in children whose mothers reported higher storm-related objective stress.	33 adolescents: 11 exposed in first trimester, 9 in second trimester, 7 in third trimester, and 6 conceived within three months after storm; no controls	Gestated during or conceived within 3 months of the storm	Degree of objective stress reported by the mother on a 32-point questionnaire of objective hardship during the storm, taken 6 months after the storm	Cytokine levels in whole-blood cultures in which an adaptive immune response is stimulated	CD3+ T cells isolated from peripheral whole blood samples	Mixed: mediation analysis of candidate CpGs sourced from whole genome study	No	No

Intergenerational Holocaust trauma

| Yehuda et al. (2016) "Holocaust Exposure Induced Intergenerational Effects on FKBP5 Methylation," *Biological Psychiatry* | Offspring had 7.7% lower methylation, on average, at 1 bin site in the FKBP5 gene compared to the offspring controls. Parent Holocaust survivors had 10% higher methylation at this site compared to the parent controls. | 32 Holocaust survivors, 22 Holocaust survivor offspring; 8 parental controls, 9 offspring controls | Preconception and/or intrauterine | Offspring: at least 1 Holocaust survivor parent and conceived after WWII (gametically or gestationally exposed to parental trauma); Parent: Jewish and either interned at a concentration camp or hidden during WWII | Differential methylation in intron 7 of the FKBP5 gene in Holocaust offspring compared to controls, and in Holocaust parents compared to controls | Peripheral whole blood cells | Candidate gene | No | Yes, but findings are not disaggregated by sex of parent |

scientists applied them to their study populations and biological samples, anticipating that epigenetics might offer a mechanism to strengthen the causal story linking exposure and outcome. All of these studies look for differential methylation levels in exposed individuals, united in the premise that these levels underlie the association between adverse prenatal exposures and later adverse effects. These foundational, highly cited studies stand as exemplars of human epigenetic findings in textbooks, scientific reviews, and writing about epigenetics for the general public.

Even prior to introducing epigenetic approaches, findings of maternal intrauterine effects were cryptic: they reported small effect sizes that varied depending on ecosocial context and that occurred at a great temporal distance from the initial exposure. Fetal epigenetic programming research conceptualizes epigenetics as cohering disparate cryptic findings of maternal-fetal effects into plausible causal stories. But as I will argue in this chapter and the next, it is not clear that epigenetic findings should strengthen causal intuitions in maternal effects science. Outstanding questions persist about the logic of the fetal epigenetic programming hypothesis, the quality of the data used to support it, and the epistemic standards that need to be met in order for epigenetic data to test hypotheses of fetal programming.

FETAL PROGRAMMING

David Barker, the British epidemiologist who founded the field of Developmental Origins of Health and Disease, or DOHaD, popularized the term "fetal programming" in the 1990s.[6] Programming, he wrote, "describes the process whereby a stimulus or insult, at a sensitive or 'critical' period of development, has lasting or lifelong significance."[7] This term was central to the operationalization of the DOHaD hypothesis, which views early life experiences as molding adult health status, and has been adopted widely: since 2000, there has been a 20-fold increase in the number of papers published on the concept, and academic textbooks and research centers around the world now feature the term in their titles.

Fetal programming researchers believe that the maternally provided prenatal environment influences development through permanent, functionally specific epigenetic modifications.[8] These researchers hypothesize that the prenatal period is critical for imprinting set-points for gene expression on the fetus's genome, which in turn modify developmental pathways in areas such as metabolism and the brain. "Even subtle disturbances of the methylation pattern," they argue, "may be associated with persisting alterations in gene expression, which may have important developmental consequences."[9] Furthermore, they believe that a growing female fetus's own gametes can be epigenetically imprinted, making the effects of the maternal environment intergenerational, passed through the maternal line to grand-offspring (see chapter 1, fig. 1.1).

Fetal programming does not simply happen—it can be something one *does*. Many fetal programming researchers embrace an applied, interventionist approach. The Danish Centre for Fetal Programming, for example, sets its aim as defining "nutritional and other lifestyle recommendations for pregnant women and . . . dietary or pharmacological interventions in early postnatal life that may prevent the adverse health outcome as a result of programming induced in fetal life."[10] Sheep researcher Steve Ford, founder of the University of Wyoming Center for the Study of Fetal Programming, hopes to use livestock studies of the effect of maternal nourishment during pregnancy on "carcass quality" to model how "fetal programming has an impact on human babies."[11] "The mother, through her diet, can modulate growth and development of the fetus," says Ford.[12] Such researchers constitute maternal-fetal epigenetic programming as a predictive science promising public health and therapeutic interventions to improve the health of future generations.

Researchers in this field conceptualize DNA methylation levels as a quantifiable molecular marker of prenatal exposures and as a causal agent of later pathology. In their words, methylation provides a "cellular memory of early life events"—that is, such events are "stably inherited" and passed from cell to cell through methylation signatures, theoretically retaining the results of early fetal exposures over a lifetime and producing biochemical modifications that lead to differences in health outcomes.[13] "Of primary interest," write researchers, "is the nature of the mechanism or mechanisms which initiate the fetal adaptations to maternal insult . . . result[ing] in permanent, irreversible physiological and metabolic changes."[14]

One hypothesis is that fetal programming is an evolved adaptation. On this view, early developmental plasticity in the uterus allows mammalian offspring to adapt their physiology to the expected postnatal environment on the basis of intrauterine cues transmitted by the mother. Peter Gluckman and Mark Hanson developed this adaptive theory in their 2005 book, *The Fetal Matrix: Evolution, Development, and Disease.*[15] "How the fetus perceives its environment is critical both to its immediate survival and also for its long-term adaptive advantage in the environment in which it anticipates living after birth," they write. In this critical process, the mother serves as a "transducer." However, "as with any electronic or mechanical transducer—for example a smoke sensor used in a fire-detection system—there is a risk of malfunction."[16] Uncovering the causes and consequences of this maternal malfunction, and redressing them, is the remit of fetal programming science.

The programming metaphor extends from a longer history of informatics metaphors in the twentieth-century molecular life sciences. The concept of the genome as a program or code originated in the mid-twentieth-century coemergence of cybernetic information discourse and scientific understanding of the structure and sequence of DNA.[17] In the most literal sense, the genetic "code" refers to the relationship between DNA sequence and amino acid synthesis worked out in the 1950s and 1960s.[18] But it also offers a broader way of thinking about the molecular basis of biology. Consider the ubiquity of computing and informatics concepts in models of the body, biology, and health embedded in our everyday language: the body as "hardware"; the genome as "software"; gene regulation as a network of "switches"; and maternal bodies as "transducers" of "signals." Even as the "genetic program," implying that the totality of instructions for an organism could be found in the genes, gives way to the notion of the "developmental program" in epigenetics and fetal programming science, the metaphor of the program, with its deterministic causal picture of goal-driven developmental signaling with highly specific adaptive functions, remains.[19]

In the fetal programming literature, scientists employ information discourse to describe how adverse exposures cause epigenetic changes, which in turn alter gene expression to cause disease. Methylation, they claim, stores information (cues from the environment) and transmits signals to reduce or increase gene output. Extending the coding/information metaphor of the genome to the epigenome, fetal programming theories imply that methylation levels, like a code, are adapted to modify gene regulation at specific sites, at specific times, and in response to specific cues. Capturing this idea, National Public Radio (US) illustrated a story on fetal programming with an image of a woman pushing switchboard buttons to activate lights on her pregnant abdomen.[20]

This view of fetal programming as a science of simple switches ready for application by savvy mothers is on display in many translations of research findings for broader audiences. Mehmet Oz, the American TV doctor popularized by Oprah, featured fetal epigenetic programming theories in his 2010 *You: Having a Baby: The Owner's Manual to a Happy and Healthy Pregnancy*.[21] "What we can learn from epigenetics is that you have the power to influence the course of biological destiny," reads the opening chapter, titled "Nice Genes." "Actions you take during your pregnancy," the book advises, "determine which genetic recipes get used . . . ultimately affecting the health of your child. . . . On, off, on, off—you decide how your child's genes are expressed." Furthermore, epigenetic changes "can be passed down from generation to generation." The upshot, the book relates in a concise popularization of the fetal programming hypothesis, is that "the environment that you provide for your offspring—through what you're eating, drinking, smoking, or stressing about—is what your child will expect of the world she's entering. Based on what you're doing, she's forecasting her future. And if the programming for gene expression doesn't match that environment, problems can occur. So your challenge—dare we say your responsibility—is to provide little Dolly with a healthy environment now so that her 'internal programming mechanism' predicts and can respond to a healthy environment later."[22]

These portrayals of epigenetics as an adaptive, causally specific mechanism that is directly manipulable—by mothers or by scientists—set up the questions of how scientists understand the biological function of methylation and how they study variation in methylation patterns in human populations. The following section may be hard-going at first for many non-scientists, but understanding how scientists conceptualize and study methylation variation in the human genome will prove worthwhile once we get deeply into the landscape of current research in this field.

AN UNSETTLED SCIENCE: THE BIOLOGY OF DNA METHYLATION AND SCIENTIFIC TECHNIQUES FOR ITS STUDY IN HUMAN POPULATIONS

Methylation, chromatin, microRNAs, histone modification, and other kinds of epigenetic agents all participate in the dynamic response of genes to regulatory signals, including those from the environment. The role of epigenetic mechanisms in basic processes of cell differentiation and organismic development has now been well documented. For instance, epigenetic factors regulated by exposure to early-life nutrition determine whether a honeybee larvae will become a queen or worker bee. Epigenetic mechanisms also play a well-studied role in the etiology and trajectory of many environmentally triggered human diseases, including cancer.

Nonhuman animal studies that allow for experimental manipulation provide the strongest evidence for the functional significance of maternal-fetal epigenetic gene expression modification. In mice with an artificially engineered gene variant linked to yellow fur color and obesity, a methyl-rich maternal diet during gestation epigenetically alters gene expression to yield brown offspring of typical body size.[23] In a species of vole, seasonal maternal melatonin levels epigenetically influence fetal coat thickness, preparing offspring for their future environment.[24] In rats, maternal licking of pups epigenetically programs glucocorticoid receptor gene expression in the brain, correlated with a low-anxiety adult phenotype.[25] In all of these classic rodent studies, the epigenetic modification is introduced via the behavior or physiology of the mother during gestation or very early development.

Fetal programming studies in human populations typically analyze one kind of epigenetic change: levels of methylation in human blood cells. High levels are generally believed to suppress gene activity, whereas low levels allow the gene to be expressed. Human studies measure individual differences in the percentage of methylation at specific loci in the human genome. Researchers then attempt to associate these quantitative differences in levels of methylation with fetal exposures and adverse outcomes.

Methylation is measured at sites in the genome called "CpGs." These sites are loci where methyl groups can attach to the DNA molecule (fig. 7.1), which is contained in the nucleus of all cells and is made up of a long string of four kinds of chemical bases—cytosine, guanine, adenine, and thymine, or C, G, A, and T, making up the genetic code. CpGs are sites where a cytosine precedes a guanine, here underlined in this short example of DNA code: AATCTGGAGA<u>CG</u>TTA<u>CG</u>ATTCAG. Epigenetic researchers can zero-in on these sites using a modification of old-fashioned sequencing methods, but more frequently epigenetic

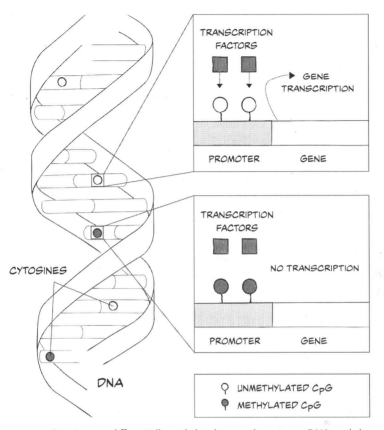

FIGURE 7.1 Cytosines are differentially methylated across the genome. DNA methyltransferase (DNMT) enzymes contribute and maintain methylation at CpG sites. Methylation at CpG sites plays a role in epigenetic regulation of gene expression. For example, methylation of CpG sites in promoter regions inhibits gene transcription. Figure by Amy Noseworthy, copyright Sarah Richardson, 2020.

epidemiologists use commercially available high-throughput "gene chips," which allow identification and quantification of variation in methylation in the human genome in large numbers of samples (fig. 7.2). Compared to sequencing, gene chips are less sensitive and specific, but they provide a much faster and lower-cost method for assessing methylation variation in epidemiological research.

There are two main approaches to identifying potential methylation mediators of associations between intrauterine exposures and later outcomes in human populations: *candidate gene* and *whole genome*. Candidate

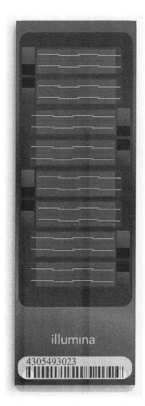

FIGURE 7.2 Illumina Infinium Methylation Microarray BeadChip. By permission from Illumina, Inc.

gene studies look at variation in methylation in a single gene, or specific loci within that gene, known to vary in methylation levels across individuals and believed to be functionally associated with the biological systems in question. The result is reported as a percentage change in levels of methylation at one genetic locus.

Whole genome studies use gene chips to rapidly assess changes in methylation levels at thousands of loci known to show variation in levels of methylation in human populations. The output of these gene chips is a numerical matrix, often represented as a fluorescent map showing the intensities of methylation at each CpG site. A value of 0 means that all copies of the CpG site in the sample are completely unmethylated, and 1 means that all copies are completely methylated. Using these data, scientists can compare values among individuals, locating any CpG sites that are differentially methylated among them. Ultimately, fetal programming scientists hope to link differentially methylated CpGs to intrauterine exposures, later health outcomes, and known functional pathways in the genome related to those health outcomes.

These "epigenome-wide association studies," or EWAS, serve as a discovery strategy, yielding findings that then must be validated using other sequencing methods as well as statistical analysis to exclude false positives. Typically, findings are reported as a list of sites, which can number in the thousands, which show significant associations with the indicator, and which are high enough in magnitude to be statistically distinguished as a significant result. The challenge for researchers is to discern whether such findings are not just statistically significant but also biologically meaningful, in the sense of delivering factors that are, in fact, causally relevant to outcomes of interest.

Because directly manipulating exposures and levels of methylation is generally not possible in human populations, a statistical technique called mediation analysis is applied by researchers to probe the plausibility of causal relationships in human epigenetic epidemiology. This method assesses whether a particular differentially methylated site could plausibly mediate (or explain a proportion of) the relationship between the exposure and the outcome. Mediation analysis is a way of modeling what might be possible, given certain facts and assumptions, and it is important to bear in mind that the statistical results are hypotheticals that must not be mistaken for biological reality. The result of mediation analyses is a set of statistical associations, not a description of actual molecular mechanisms explaining the phenotypic states of individuals.[26] Especially important for the discussion to follow, mediation analysis makes strong and often controversial assumptions about the integrity of measurements of the outcome and the likelihood of unmeasured confounders.[27] If mediators are confounded—that is, another variable is associated with both the exposure and outcome—then the analysis is invalid. Researchers agree that unmeasured confounding is highly likely in complex epidemiological data sets such as those in use in human environmental epigenetics.[28]

One signal of the unsettled state of the epigenetics knowledge base is that there is little consensus in the literature as to what constitutes a functionally significant, reliable, and valid magnitude of change in levels of methylation. Two facts are well recognized. First, methylation usually regulates genes by completely methylating them. Methylation state is generally bimodal—either methylated, or not. Thus, the gene chips feature only a small subset of loci that are not bimodal and vary in levels of methylation from individual to individual. Second, within this subset of CpGs that show interindividual variation in methylation, there is a lot of background noise. Many variables can cause changes in methylation levels—not just environmental exposures during critical developmental periods. Methylation levels may vary over the course of the day, differ between sexes and across populations, and change as people age. Even season and weather have been shown to influence human DNA methylation variation.[29]

Different methods of detection also produce varied results. The Infinium Human Methylation 450k Beadchip microarray, marketed by the San Diego biotech Illumina, is widely used in whole genome methylation studies. Epigenome-wide association studies using the Illumina array carry inherent technological and statistical limitations. This approach analyzes a large number of relationships that nonetheless represent only a minor subset of CpGs in the genome. The current version of the Illumina microarray includes fewer than 2 percent of the CpG sites in the human genome. Moreover, the Illumina array has low specificity for any individual CpG site, because it assumes that CpG sites located adjacent to those picked up by the probe have a similar methylation state. The array is also far less reliable than sequencing methods for detecting small changes in methylation levels that characterize human studies.[30] The design of the chip is also not hypothesis-neutral. While the chip includes CpGs located in most of the functional regions of the genome identified by the Human Genome Project's reference sequence, it preferentially targets known regions of variability in regulatory regions deemed important by leading researchers in the field who were consulted in the chip design process.[31]

Many nonhuman animal studies report methylation differences only when they are greater than 10 percent.[32] Foundational animal studies demonstrating functional consequences of methylation variation suggest that changes of greater than 10 percent are required to produce meaningful functional phenotypic changes. In studies of how rat maternal licking and grooming of offspring alters methylation of the glucocorticoid receptor promoter, determining stress phenotype, low- compared to high-licking and grooming offspring are nearly completely unmethylated or wholly methylated.[33] In the well-known study of the role of diet in determining methylation-regulated expression of coat color in mice, yellow mice are less than 10 percent methylated, whereas brown mice are typically 90 percent methylated at the site in question. In-between phenotypes of slightly mottled, mottled, and heavily mottled correlate with methylation levels around 20, 40, and 60 percent respectively.[34] In contrast, most human epigenetic studies in fetal programming science report methylation differences well below 10 percent, findings that would typically not be considered functionally meaningful in nonhuman animal models.

A central area of dispute concerns whether the epigenetic differences observed in a single human sample—of, say, blood—is a reliable guide to methylation levels in the biologically relevant tissues for the research question, such as immune, brain, metabolic, endocrine, or adipose tissue. A 2012 study demonstrated that peripheral tissues such as whole blood do not well model the epigenomic state of other bodily tissues. If, as this research suggests, the epigenome is tissue-specific, then studies using human blood samples cannot, without validation, serve as the basis for hypotheses regarding epigenetic regulation in health conditions involving other organs.[35] Nonetheless, blood is routinely used for human fetal epigenetic programming studies.

Moreover, epigenetic studies that use blood samples assess a mixture of cell types that characterize blood tissue, creating many possible confounding effects in measurement. Blood is itself a heterogeneous mix of cell types, and the heterogeneity of its composition varies among individuals. As a 2014 *Lancet* article noted, because of this variation, "if the cellular content of a tissue is strongly associated with the trait being studied—as is likely for inflammatory or neurodegenerative disorders, for example—any apparent trait-associated epigenetic differences could partially reflect differences in cellular composition, even after statistical correction."[36]

Relatedly, many discoveries of DNA methylation differences can be explained simply by genetic differences between people. An individual's genes determine much of their distinct epigenetic profile—even as much as 80 percent of it, according to one study.[37] Cell type composition and individual variation in background DNA are just two of many potential influences on epigenetic variation. Without a thorough understanding of the endogenous and exogenous factors that influence changes and set-points in human epigenetic gene regulation, many studies reporting variation in methylation levels and associating this variation with environmental exposures are difficult, if not impossible, to interpret.[38]

STUDYING FETAL EPIGENETIC PROGRAMMING IN HUMAN POPULATIONS

Project Ice Storm

McGill University developmental psychologist Suzanne King saw a unique research opportunity when a 1998 ice storm sent the nearby Montérégie region of Quebec into weeks of hardship. With the power out and roads blocked, three million people suffered injuries, extreme cold, and financial and property losses for up to six weeks. "I knew there were thousands of pregnant women out there. I could watch it happen," King recalls.[39]

Within months, King initiated a longitudinal study to track babies gestated during the storm until they reached the age of 18. She recruited 224 new mothers, who completed a mail-in questionnaire of objective hardship (such as injuries, days without power, and economic loss) and subjective distress (mothers' recall of how stressed they felt) during the storm. At regular intervals in the following years, "Project Ice Storm" re-surveyed 176 of these mothers and conducted physical exams on their children.

To date, Project Ice Storm has produced 30 peer-reviewed scientific papers, in publications such as the *Journal of Developmental Origins of Health and Disease, Pediatric Research,* and *Epigenetics,* advancing the fetal programming hypothesis. Early results found associations between the degree of objective hardship that the mother experienced during the storm and many diverse outcomes, including asthma, autism traits, birth defects, obesity, body mass index (BMI), central adiposity, insulin secretion, cognitive and linguistic functioning, immune profile, age of menarche, and eating disorders. "One of the scariest studies," *Newsweek* reported, revealed differences in play behavior.[40] Instead of "using the toy truck as a truck," a child of a prenatally stressed mother was described as "simply banging around the object randomly." Couching the results in grade-point averages, King translated this behavior as the difference between "a child who is going to average B-minus as opposed to a child who will get B-plus."[41]

As the children reached age 13, King added epigenetics methods to Project Ice Storm's toolkit. King extracted immune cells from the blood samples of 34 children, seeking epigenetic markers that might explain how exposure to maternal stress during gestation relates to the cognitive, metabolic, immune, and hormonal outcomes she observed in adolescents. Using the Infinium Human Methylation 450k Beadchip microarray, her team examined nearly half a million sites in the human genome, assessing epigenetic variation among teenagers whose mothers experienced higher compared to lower objective hardship during the ice storm. Then, using reference databases, her team examined whether these methylation changes occurred in biological pathways that might be related to the many health and developmental outcomes that Project Ice Storm had discerned over the years. The results, published in 2014, revealed that adolescents with the most storm-stressed mothers had differential methylation levels at 1,675 sites in the genome in immune, neurological, and endocrine pathways. The paper described this as the "first evidence in humans supporting the conclusion that prenatal maternal stress results in a lasting, broad, and functionally organized DNA methylation signature . . . in offspring."[42]

The media response to these findings was significant. One *London Times* headline on King's research read, "Feeling stressed? Then blame your mother; Your well-being is determined by how relaxed she was . . . when she was pregnant with you."[43] Maternal stress during pregnancy "significantly stalls children's development, making them less intelligent, more anxious, and more prone to bad behavior," reported the *National Post*.[44] "It's a new insight for soon-to-be moms," wrote the *Global News*. Once maternal stress is "programmed into" children, the article advised, nothing in their life experience can overcome the long-term implications.[45]

Since the 2014 study, King's team has published several further analyses of whole genome methylation results in thirteen-year-olds conceived or gestated during the 1998 ice storm. One study examined how a mother's cognitive appraisal of ice storm stress—that is, whether the stressor was perceived in a positive or negative light by the mother—affected offspring DNA methylation. This revealed 2,872 differentially methylated sites in the affected offspring, featuring many genes and biological pathways different than those found in the previous study of offspring methylation and the mother's objective degree of hardship.[46]

From study design to the interpretation of data, the fetal programming hypothesis explicitly frames the Project Ice Storm research program. King represents methylation as a "mechanism" that mediates the relationship between prenatal maternal stress and "programming" of "poor health outcomes later in life," including cardiovascular disease, obesity and metabolic syndrome, and immunity.[47] Specifically, King and colleagues argue that "poor outcomes in the offspring" are caused by "system wide" but functionally organized "epigenetic modification of gene function" that may "persist throughout life."[48]

Researchers point to King's work as a model for the promise of epigenetic methods for examining the health effects of prenatal maternal stressors.[49] The ice storm is said to offer a "natural experiment," a rare opportunity to study the effects of prenatal exposure to maternal stress in a human population in which exposure to the stressor was randomized. King argues that Project Ice Storm findings have broad ramifications. For example, with the rise of natural weather disasters due to climate change, King has suggested that Project Ice Storm's findings can support public health researchers in understanding how to better shield pregnant women from stress during these events and support the infants developmentally affected by them. Canadian climatologist Gordon McBean, a member of the Intergovernmental Panel on Climate Change (recipient of the 2007 Nobel Peace Prize), has even stated that he hopes King's research might "spur countries . . . to slash carbon emissions."[50] Citing Project Ice Storm, climate activists warn of an approaching wave of "climate change babies" bearing the imprint of prenatal exposure to weather disasters and presenting "a wide range of health problems."[51]

The longitudinal design in a natural population, with many repeat measures, is a strength of Project Ice Storm. Nonetheless, there are several limitations to its epigenetic findings. Typical of much human epigenetics research on prenatal influences, the sample size of 34 is small. So, too, are the effect sizes. All children were in the normal and healthy range for outcomes. For instance, although the children of mothers who experienced higher objective stress during the storm scored, on average, a bit higher on a scale of "autism traits," they did not have autism. And, unfortunately, Project Ice Storm does not include a control group—that is, a group of matched children and mothers who did not experience the storm. Instead, King uses the degree of stress experienced by mothers to stratify the group and analyze variation among children. Because the studies in this research stream examine the dose-response relationship between maternal stress and a variety of outcomes entirely within the exposed group, the sensitivity and validity of the five-point scales on the questionnaire used to assess degree of maternal objective and subjective stress during the storm is extremely important to interpreting their findings. Furthermore, without the inclusion of data on fathers (and other conceptional, prenatal, and perinatal influences), these studies cannot localize causality specifically to maternal intrauterine exposures. Additionally, while prior literature in the field of fetal programming science strongly suggests that prenatal influences are timing-specific, and generally larger when experienced at periconception, Project Ice Storm includes a wide timespan of maternal prenatal exposure, complicating the causal story. Children either gestated during the storm or conceived within three months after it—a period of 17–21 months over the course of two years—are included without stratifying the results by timing of exposure.

Finally, Project Ice Storm did not collect biological material at birth. This is a problem for two reasons: reverse causation and confounding. Reverse causation can occur when the outcome condition—for example, obesity—*itself* causes epigenetic changes. In that case, only a time-sequence demonstrating the induction of an epigenetic change relevant to obesity at an earlier time point can support a causal narrative leading from epigenetic change to the condition, and not the other way around.

The ideal study design for researching the role of epigenetics in associations between intrauterine exposures and long-term health outcomes is a longitudinal birth cohort, with multiple biological measures and clinical assessments in both parents and offspring, starting at conception or during the prenatal period and continuing through adulthood. This design would develop a time-sequence that could help address reverse causation. Presently, few studies in fetal origins research meet this criterion, in part because such study designs are costly and can require decades or more to yield useful data. Only in the late 1980s and 1990s did epidemiologists begin to track exposures from conception onwards, in maternal-fetal dyads, and many studies are considerably more recent than that. Without this sort of data, Project Ice Storm cannot rule out reverse causation—that is, that the methylation changes they observed may be caused by higher body fat at age 13.

This is related to the second issue, confounding. Only a temporal sequence of exposure, epigenetic marks, and outcome, in that order, is consistent with the hypothesis that methylation operates as a biological mechanism mediating an association between a fetal exposure and an outcome. But even when such a sequence exists, confounding factors could account for the relationship. For instance, a finding that maternal obesity during pregnancy is associated with later metabolic problems in offspring could be because of the direct effect of gestation in an over-weight mother, or because the offspring adopted a similar lifestyle as the parent after birth, among other possibilities. In the case of Project Ice Storm, maternal stress may not stop at birth, but could continue to developmentally impact children postnatally, creating a confounding effect. As a result, the study cannot confirm that the associations between *fetal* exposure and the outcomes are causal.

When King first applied epigenetics approaches in 2014, she hypothesized that methylation changes associated with higher stress "may be responsible" for the health effects observed in children of mothers who experience prenatal stress. That is, King's group believed that differentially methylated sites in the genome could explain some of the biological outcomes, such as higher levels of BMI, obesity, insulin secretion, and immune response observed in the children of women who endured the most stress during the storm. But when King's team employed mediation analysis in two 2015 and 2016 studies, that is not what they found. Instead, results suggested the opposite, that the methylation changes appeared, if anything, to have "dampened" the effects of prenatal maternal stress on outcomes such as increased BMI and excessive immune response.[52] Unswervingly committed to a causal, functional hypothesis of the role of variation in methylation levels in mediating associations between fetal adversity and offspring outcomes, in response to this finding, King and her colleagues reframed their hypothesis. Rather than *causing* adverse outcomes, methylation changes found in 13-year-old intrauterine survivors of higher gestational stress during the ice storm are now seen as a protective "response to early adverse environmental factors."[53]

Any such causal interpretations of the methylation changes found in the Project Ice Storm adolescents, however, require enormous inferential leaps given limitations in research design and in knowledge about the functional specificity of methylation changes in human maternal-fetal epigenetic programming studies. Project Ice Storm has now concluded with no further plans for follow-up. A larger epigenome-wide study of 1,740 cord blood samples from neonates has subsequently found no association between prenatal maternal stress exposure and differential DNA methylation.[54] While the result may be due to use of different biological materials and the inclusion of a range of stressors rather than one set of hardships, such as the ice storm, such a discordance makes clear that there is much yet to understand about the role, if any, of epigenetics in mediating posited effects of prenatal and preconceptional maternal stress on a child's epigenome.

Intergenerational Inheritance of Holocaust Trauma

Neurogeneticist Rachel Yehuda's 2016 study of Holocaust survivors and their offspring stands among the highest-profile scientific claims to date positing a persistent epigenetic marker in adult humans associated with early-life exposures.[55] The study sought an epigenetic mechanism to explain earlier work by Yehuda showing that Holocaust survivors in her sample had higher cortisol, a critical hormone in stress regulation implicated in the development of PTSD. By contrast, on average, their children had lower cortisol than the general population.

Yehuda hypothesized that lower cortisol levels in the second generation could be the result of intrauterine environment. Bathing in high cortisol levels in the womb might produce a protective over-response in the fetus, leading to greater glucocorticoid sensitivity later on.[56] Permanently set in place by epigenetic markers, this sensitivity, Yehuda suggested, could render offspring more vulnerable to psychiatric disorders throughout their lives, helping to explain intergenerational patterns in anxiety and psychiatric disorders observed in the descendants of trauma victims.

For the 2016 study, Yehuda undertook to assess epigenetic variation in Holocaust survivors and their children at one genetic locus, a noncoding

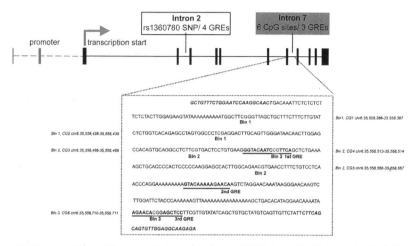

FIGURE 7.3A Schematic representation of the human FKBP5 locus highlighting the six CpGs across three "bins" in intron 7 investigated by Yehuda et al. in 2016. GREs are "glucocorticoid response elements." From Yehuda et al., "Holocaust Exposure Induced Intergenerational Effects on FKBP5 Methylation." By permission of Elsevier.

segment of DNA known as intron 7 of the FKBP5 gene. As shown in fig. 7.3a, introns are segments of DNA interspersed within the protein-coding portions of the genome. Introns often contain regulatory elements important to gene activation. A 2013 paper, by Klengel et al., had reported lower levels of methylation at intron 7 among humans exposed to high levels of childhood abuse.[57] Hence, Yehuda considered it a promising potential site for examining the epigenetic imprint of Holocaust survival.

Using blood samples from North American Jewish Holocaust survivors, their children, and a group of demographically and ethnically matched control subjects from families that did not experience the Holocaust, Yehuda's team scoured FKBP5 intron 7 for changes in methylation levels. At one site, they found methylation levels 10 percent higher in Holocaust survivors, but 7.7 percent lower in their adult offspring, compared to controls. This finding, they wrote, is "the first demonstration of an association of preconception stress effects with epigenetic changes in both exposed parents and their offspring in adult humans." They concluded that "our data support an intergenerational epigenetic priming of the physiological response to stress in offspring of highly traumatized individuals."[58]

Situating the maternal-fetal interface at the nexus of the intergenerational transfer of trauma, Yehuda's research has been embraced across a wide range of academic fields and social arenas. Psychologist Nirit Gradwohl Pisano, author of a book on third-generation Holocaust survivors, *Granddaughters of the Holocaust: Never Forgetting What They Didn't Experience*, cites Yehuda in making the claim that "frequently unspoken, unspeakable events are inevitably transmitted to, and imprinted upon, succeeding generations."[59] Natan Kellermann, of the National Israeli Center for Psychosocial Support of Survivors of the Holocaust and the Second Generation, argues that Yehuda's epigenetic research may provide "verifiable data of trauma transmission," preparing the way for legal reparations for "epigenetically inflicted wounds."[60] "The Holocaust left its visible and invisible marks not only on the survivors . . . but also on their children. Instead of numbers tattooed on their forearms, however, they may have been marked epigenetically with a chemical coating upon their chromosomes," he writes.[61]

Poet Elizabeth Rosner, drawn to the promise of epigenetics as an embodied mechanism of memory for Holocaust descendants, placed Yehuda's work at the center of her 2017 memoir reflecting on the imminent passing of the last living Holocaust victims. "During the next few years, certainly within another decade as the final survivors of World War II die off, we will all experience the threshold where the Holocaust becomes disembodied history," she wrote. For Rosner, Yehuda's suggestion that the past is "transmitted through our cells, our DNA," offers one hopeful answer to "how we will remember what happened to an entire generation once they are no longer here to participate in the conversation." Right now, the still-living survivors represent "embodied history, bearing witness," but perhaps, she suggests, epigenetics will allow second- and third-generation descendants to also be embodied survivors: "We will continue upholding our promise to pass the story forward, generation to generation, as though each of us were there. We will embody the DNA of the dead."[62]

For Pisano, Kellermann, and Rosner, epigenetic marks on the body and biology help keep alive memories, and can serve as evidence that the crime of the Holocaust, and its ensuing trauma, was real. "Every time a new study proves the epigenetic effect of severe trauma," writes Rosner, "the underlying fact of my inheritance gets its scientific affirmation, especially in a world where measurable evidence is always preferable to implication." By contrast, mere language (and the stories it is used to share) has its "limits . . . its inadequacies and inaccuracies." Biological or bodily evidence brings "real" proof to longstanding claims by survivors and their descendants. "The research is reassuring and unsettling, offering an intricate mixture of relief and resignation. If childhood starvation can change your metabolism forever, maybe the body's memory is just as persistent as the mind's. Maybe more."[63]

Among the research streams analyzed in this chapter, Yehuda's is perhaps the most audacious in asserting the causal specificity and stability of a particular marker. Yehuda believes that methylation changes at intron 7 of the FKBP5 gene represent an adaptive priming of stress response in a functionally organized pathway. Methylation changes at this site, she claims, reflect a specific kind of trauma exposure and are stable across adulthood. Alterations at this specific site can also be transmitted chemically by exposure to stress hormones in the womb. Further, Yehuda posits that methylation changes at this site contribute to an individual's psychiatric risk profile and represent a manipulable target for interventions to change mental health outcomes.

Yehuda, who holds patents for "Genes associated with post-traumatic stress disorder" and "FKBP5: A novel target for antidepressant therapy," has described her 2016 finding as part of ongoing work to develop "treatments to target epigenetic abnormalities" among those exposed to trauma.[64] A *New Republic* article titled "The Science of Suffering: Kids Are Inheriting Their Parents' Trauma—Can Science Stop It?" profiled Yehuda as seeking "to locate the exact spots on genes where molecular changes occur in response to trauma" with the hope of identifying precise "reversible . . . targets" for medical intervention.[65] A future "epigenetic medicine" that builds on findings such as Yehuda's, wrote one commentator in a psychiatry journal, might allow for more "accurate diagnosis" and "targeted treatment" of trauma and PTSD, including through the use of psycho-pharmacological drugs.[66]

But is it biologically plausible that methylation variation at a single site in intron 7 of the FKBP5 gene could cause increased stress reactivity in humans? Yehuda grounds the case for epigenetic changes at this site as a cause of "increased risk for psychopathology" in the 2013 work done by Klengel et al., mentioned above.[67] However, this is a shaky foundation for Yehuda's functional inferences.

In Klengel et al.'s study of adult victims of childhood abuse, extreme childhood abuse coupled with epigenetic changes at three sites in the FKBP5 gene, in the presence of a particularly risky variant of the FKBP5 gene, were linked with a higher likelihood of developing psychiatric disorders after trauma. Klengel et al.'s 2013 findings were larger, at different sites, and in a different direction than those in the 2016 Yehuda paper. Klengel et al. found a 12.3 percent *decrease* in methylation at three loci (in intron 7, bin 2, sites 3 and 5). Yehuda et al. found opposite effects in parents and offspring at a different site (intron 7, bin 3, site 6), and no differences across all subjects and controls at other sites in intron 7. And although Klengel et al. observed a decrease in methylation in intron 7 among victims of trauma, Yehuda found an *increase* of methylation at this site in adult victims of the Holocaust.

Klengel et al.'s findings were different in another crucial way: inherited genes also mattered. The decrease in methylation that Klengel's team observed was only present in child abuse survivors who were also carriers of a genotype with polymorphisms associated with risk of psychiatric disorders after trauma. In the Klengel study, FKBP5 was differentially methylated in those who had directly experienced childhood abuse, but this change in methylation was linked to later PTSD only in the presence of an inherited gene variant. By contrast, experience of abuse was not an exposure and PTSD was not an outcome in the Yehuda study. In contrast to the Klengel study, in Yehuda's study, an epigenetic effect was present in the Holocaust survivor parents and in the offspring even if they themselves did not show evidence of PTSD, and the effect was unmodified by presence or absence of the risky gene variant.[68] Only the experience of surviving the Holocaust, in the parents, and being the child of such a parent, in the offspring, correlated with epigenetic alterations at this site (albeit in different directions for parent and offspring).

But is there in fact a stable signature of the Holocaust at this site that is distinct to familial Holocaust exposure? The answer is not clear from Yehuda's data. As fig. 7.3b, reproduced from Yehuda's 2016 paper, shows, both control and Holocaust parents and offspring, as well as matched controls, had similar levels of methylation at two sites examined by Yehuda's team. The average methylation level at the third locus is slightly lower—by 3 points—in Holocaust offspring than in their own parents. In absolute terms, Holocaust survivors' level of methylation

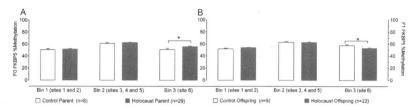

FIGURE 7.3B Methylation at FKBP5 intron 7, bins 1, 2, and 3 (A) for Holocaust survivors, (B) for Holocaust survivor offspring, compared to their respective controls. Only bin 3, site 6 shows a statistically significant difference, but in different directions for parents and offspring. The largest difference is between the parent and offspring control groups. From Yehuda et al., "Holocaust Exposure Induced Intergenerational Effects on FKBP5 Methylation." By permission of Elsevier.

was 5 percentage points (56.39 vs. 51.33) higher than controls, while their offspring were 4.45 points (53.13 vs. to 57.58) lower than controls. That is, the largest change in methylation levels that the study found between generations is among the non-Holocaust-exposed control parents and offspring, not between controls and the Holocaust families. In the non-exposed controls, offspring methylation levels were 6.25 percentage points higher than their parents. It appears that in the controls used in this study, the average methylation levels jumped *up* (6.25 percentage points) in the offspring compared to the parents more than they *declined* (3 percentage points) in the offspring of Holocaust parents. Rather than a strong effect in the offspring of Holocaust survivors, the central finding of the study—10 percent higher methylation in Holocaust survivors, but 7.7 percent lower in their adult offspring—may be an artifact of a strong effect in the controls.

Yehuda utilized a tiny control sample of 8 parents and 9 offspring, compared to 32 parents and 22 offspring in the Holocaust-exposed group. With a small number of controls, a single outlier can easily distort results. Indeed, as fig. 7.3c, reprinted from Yehuda's paper, illustrates, there was a great range of variation among the Holocaust parent-offspring pairs and an extreme outlier in the control offspring group. Average methylation levels across all groups ranged from 40 to 70 percent. While the mean Holocaust parental methylation level was slightly higher than their offspring, individual parent-offspring pairs exhibit an overall positive correlation between levels of methylation at the FKBP5 intron 7 site: that is, as the parents' level of methylation went up, so, too, did their children's.

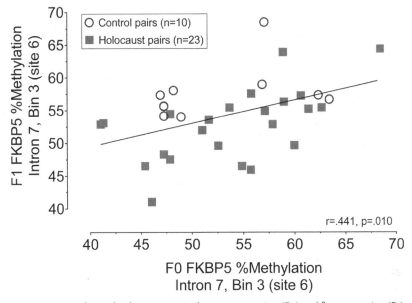

FIGURE 7.3C Relationship between original parent generation (F0) and first generation (F1) *FKBP5* intron 7 bin 3/site 6 percent methylation. From Yehuda et al., "Holocaust Exposure Induced Intergenerational Effects on FKBP5 Methylation." By permission of Elsevier.

The very highest level—70 percent—of methylation at the FKBP5 gene site in question recorded among all 71 individuals sequenced in the study was found in a single individual among the control offspring. Thus, a skew among the control offspring may very well have driven the study's findings.

Even were the finding to be replicated in larger studies, it would be challenging to ground a causal biological story in Yehuda's evidence of a few percentage points of change in methylation levels at intron 7, bin 3, site 6 of the FKBP5 gene. As geneticist Evan Boyle has pointed out, with, for example, more than 100,000 genetic polymorphisms associated with human height, and evidence that "even the most important loci in the genome have small effect sizes," causal stories about the role of individual genetic (or gene regulatory) factors can be established only by experimentally probing how risk variants affect biological response "within narrowly defined cell types or under precise conditions such as immune stimulation."[69] Combined with the challenges of replicating epigenetic research findings across diverse populations and data sets, the complexity and context-specificity of genome regulation represent a profound challenge to causal claims in epigenetic science, such as Yehuda's, based on findings of small variations in methylation levels at a few, or a single, CpG site in the human genome.

The Scars of the Dutch Famine, Six Decades On

After a hiatus of several decades, interest in the cohort of infants gestated during the Dutch Hunger Winter rekindled in the 1990s, when obesity and related metabolic conditions, including high blood pressure, diabetes and insulin resistance, and cardiovascular disease, became widely perceived as a public health emergency. A neglected finding of the original 1970s Dutch Hunger Winter research (described in chapter 6) came to light: a 50 percent higher rate of obesity among 18-year-old male military conscripts who had been gestated during the 1944 famine.[70] This finding now appeared intriguingly concordant with David Barker's prediction that when prenatal nutritional deprivation later meets nutritional overabundance, the "mismatch" in prenatal programming and developmental environment can lead to metabolic dysregulation.[71]

By the 1990s, gestational Dutch Hunger Winter survivors were now entering their fifties, a prime period for the emergence of the sorts of chronic diseases increasingly burdening national budgets. Through intensive outreach efforts in collaboration with the Dutch health system, a new generation of researchers developed expanded databases of individuals gestated during the famine. Hoping to explore whether Dutch Hunger Winter survivors had anything to reveal about the subtle longer-term effects of prenatal undernutrition, researchers interviewed them, conducted physical examinations, retrieved birth records, and took blood samples. Among the noted findings: on average, survivors had elevated blood pressure, and women gestated during the Dutch Famine weighed a bit more, at age 59, than their non-exposed siblings.[72]

The Dutch Hunger Winter Families Study was initiated by L. H. Lumey in the early 1990s. It began with 3,307 exposed and unexposed control individuals (of similar ages, and in some cases, siblings) across three Dutch cities. In 1992, they were traced to their current addresses and interviewed. Between 2003 and 2005, at age 59, they underwent physical examinations, including a blood draw.

These blood samples became the material for the breakthrough 2008 epigenetic study described in the introduction to this chapter. Seeking to detect distinctive epigenetic changes in those gestationally exposed to famine, in 2008 Heijmans and Lumey focused on IGF2, perhaps the best-characterized human gene under epigenetic regulation. IGF2, or "insulin-like growth factor 2," is a key mammalian gene in placental and fetal growth. Humans receive two copies of the gene, one from each parent, but the maternal copy is silenced by methylation. When methylation preventing expression of the maternal copy of IGF2 is disrupted, disease results. Beckwith-Wiedemann Syndrome is a dramatic example of this—the IGF2 gene is incompletely silenced in infants with this syndrome, and they are born dramatically overgrown.

Heijmans and Lumey found that individuals exposed to the famine had 5.2 percent lower methylation across several sites in the IGF2 gene compared to their unexposed siblings. For context, this represents a difference in methylation levels at the IGF2 locus just slightly greater than that between two individuals with an age gap of 10 years, which yields a 3.6 percent decline in IGF2 methylation.[73] Heijmans and Lumey described these findings as evidence supporting the hypothesis that epigenetic mechanisms underlie the relationship between early life exposures and later metabolic disease.[74]

The authors were clear about the limited conclusions that could be reasonably drawn from this finding. First, the results show correlations only between famine exposure and methylation levels—the study did not examine the functional relationship between methylation levels and metabolic disease. Second, the study could not demonstrate that maternal diet, in particular, creates persistent epigenetic changes. Without measures of epigenetic markers in the fetus or infant at birth, the hypothesis that these changes were induced gestationally and then persisted throughout life could not be validated. Even were it possible to locate the induction of the methylation changes to the fetal period, it would be impossible to distinguish whether diet or other factors, including disease, stress, and toxin exposures associated with wartime and famine conditions, are the appropriate culprit.

The 2008 study represents a single-gene, opportunistic approach to the study of epigenetic variation caused by fetal exposures, characteristic of a first wave of human epigenetics explorations in fetal origins research.

Heijmans and Lumey well recognized that a finding of a small epigenetic difference, in a relatively small population sample, at this single locus, in whole blood tissue, cannot substantiate a causal biological mechanism without a stronger scientific understanding of the relationship between slight changes in levels of methylation at this site in the genome in a blood sample, other biological and environmental factors, methylation variation in metabolically relevant tissues and organs, and human health outcomes.

Since 2008, Heijmans and Lumey have employed a variety of methods to further explore the possibility that epigenetic mechanisms might mediate the long-term metabolic effects of famine exposure in utero. In one study, they looked at whether normal background genetic variation contributes to variation in methylation levels at the IGF2 locus. The results showed that individual DNA variation can produce the same half-standard-deviation shift in methylation levels at IGF2 as they found in the 2008 study correlating famine exposure with methylation changes. Thus, epigenetic changes of the sort observed in the 2008 study can be caused by genetic differences instead of, or in combination with, the hypothesized prenatal environmental exposures.[75]

In two papers published in 2015 and 2018, the research group applied the whole genome method used by Project Ice Storm to their sample of 59-year-old Dutch famine survivors, now expanded to a group of 348 survivors and 463 time-matched or sibling controls.[76] Surveying more than a million differentially methylated CpG sites in the genome, they found four loci exclusively associated with the experience of famine during the first ten weeks of gestation. The changes were small: the four CpGs showed a 2.3 percent increase, 2.3 percent decrease, 0.7 percent increase, and 2.7 percent increase in methylation.

In 2018, Heijmans and Lumey used the same whole genome data set as the 2015 study to engage a mediation analysis to locate differences that might plausibly causally mediate metabolic outcomes such as higher BMI (one measure of obesity) and triglyceride levels. This yielded nine *entirely different* loci with altered levels of methylation in gestational famine survivors than those identified in 2015. The initial target, the IGF2 locus, did not turn up as a mediator of metabolic outcomes in any of the whole genome studies.

Mediation analysis suggested that one CpG explained about 13.4 percent of the association between prenatal famine exposure and BMI, and that together the other six CpGs explained 80 percent of the association between prenatal famine and triglyceride levels. Two other CpGs explained associations distinct to very early prenatal famine exposure. "Our data are consistent with the hypothesis that epigenetic mechanisms mediate the influence of transient adverse environmental factors in early life on long-term metabolic health," Heijmans and Lumey wrote. But they highlighted that the nine CpGs identified as mediators "are not known to play direct roles in fat and TG [triglyceride] metabolism" and "are unlikely to be directly involved in a mechanistic sense" in causing these outcomes.[77] Instead, they posited that they may be features of broader "adverse morphological or cellular" changes.[78] "The *specific* mechanism awaits elucidation," they concluded.[79]

ARE EPIGENETIC FINDINGS PROOF THAT MATERNAL INTRAUTERINE EFFECTS ARE REAL?

Project Ice Storm, Yehuda's intergenerational Holocaust trauma research, and the Dutch Famine studies are bold research programs seeking to validate the fetal programming hypothesis using imperfect human samples and emerging technologies. As this review has demonstrated, proper interpretation of their findings against the backdrop of the limitations of the study design and the technologies for assessing methylation differences themselves is key.

Heijmans and Lumey's exploration of potential epigenetic mediators of patterns identified among survivors of gestational exposure to the Dutch Famine is distinctive within the group of studies analyzed in this chapter, not only for its relative humility but also for the rigor of study design. If causal stories are going to be told about epigenetic mechanisms in developmental origins of health and disease, it seems likely that they will look like this one.

Sample size is critical to the strength of research findings. The smaller the sample, the more likely that a small, arbitrary skew in the sample population can alter the results. Notably, the 2018 study includes 348 individuals gestationally exposed to the famine. By comparison, Project Ice Storm and Yehuda's study included just a few dozen individuals in their exposed cohorts. Although 348 is still a tiny sample in the world of biomedicine, at more than 10 times the size of these other studies, it holds more potential for identifying generalizable patterns in associations. The Dutch Famine studies also include well-matched controls, utilizing a mix of same-sex siblings of the famine survivors and time-matched individuals with a similar demographic profile as the exposed individuals. Furthermore, these studies feature well-specified exposures. Although there are acknowledged study limitations due to the retrospective nature of reconstructing the cohort decades after the exposure, scientists know to a remarkable degree of precision the number of days during which communities were exposed to famine, and they also know the dosage of the exposure, a result of well-kept records of caloric allocations across the famine. Importantly for fetal origins research, where strong evidence supports the special importance and magnified effects of very early periconceptional exposures as well as distinctive outcomes depending on first or last trimester exposure, this allows researchers to specify the exposure to particular time windows during pregnancy and assess the relationship between time of exposure, epigenetic changes, and outcomes. Outcome measures are also well defined and hypothesis-driven. Rather than taking a broad discovery approach to finding associations between any epigenetic variation and any outcome measure, as in Project Ice Storm, Heijmans and Lumey examine epigenetic markers in pathways related to growth and metabolism and to outcome measures, such as blood pressure, BMI, and triglyceride level.

Despite these methodological strengths, there are serious doubts about Heijmans and Lumey's findings. Different CpGs were found in all three study approaches, including in the two that used the same whole genome data set. None of these epigenetic markers have themselves been replicated as mediators of the effects of famine exposure using different methodologies or different cohorts. And several critics have since offered compelling evidence that confounding or reverse causation is the more likely explanation for Heijmans and Lumey's results.

Subsequent statistical reanalysis of Heijmans and Lumey's 2018 study by Bristol University epigenetic epidemiologists found "no convincing evidence for a causal effect of methylation on triglycerides." "Unfortunately, the very sophistication of the methods—and their impenetrability to many—can generate a false sense of security regarding the results obtained," they wrote.[80] Mediation analysis in human epigenetic research on maternal intrauterine effects, the critics argued, is "beset with methodological problems," which are "highlighted" by Heijmans and Lumey's 2018 paper. Pointing to evidence suggesting that elevated triglycerides (TG) could cause the epigenetic changes found by Lumey and Heijmans (rather than the other way around), Albert Einstein College of Medicine scientist John Greally tweeted in response to the 2018 findings, "So if TG—>DNA methylation, should the mediation analysis be interpreted to mean that high triglycerides cause famine earlier in life? . . . mediated by DNA methylation? The power of #epigenetics. Amazing. It doesn't just alter your future, it alters your past . . . Editors, reviewers, journals, funding agencies: please start to act responsibly and enforce rigour. #epigenetics is science, not magic."[81]

The question of whether maternal-fetal epigenetic programming is "magic" or science takes us full circle to the debates, discussed in the opening chapters of this book, ignited more than a century ago between the Weismannians and prenatal culturists. Chapter 8 explores how researchers today grapple with the problem of cryptic causality in the science of maternal intrauterine effects.

CHAPTER 8

It's the Mother!

The rise of Developmental Origins of Health and Disease and the fetal programming hypothesis is part of a forceful reassertion, over the past decade, of wide-ranging theories of the maternal-fetal interface as a critical determinant of lifelong health and of intergenerational patterns in disease distribution. The sudden and remarkable intensity of interest in fetal programming today recalls the efflorescence of prenatal culture theories of the relationship between pregnancy, health, and heredity in the late 1800s and early 1900s United States.

Foundational to both moments is the belief that the fetal period is characterized by a high level of plasticity in response to environmental input, that early environmental exposures leave an imprint that is relatively fixed and stable for life, and that the imprint left by intrauterine environmental factors leaves a greater legacy than do other contemporaneous and postnatal environmental inputs. In both eras, theories emphasize the distinct importance of maternal prenatal factors relative to other early developmental variables, including those related to the father and the wider social environment. As in eugenic-era prenatal culture theories, DOHaD today conceptualizes the overall constitutional health of the parent as conferring a somatic imprint on the offspring with potential intergenerational consequences.

In both eras, as well, researchers wrestled with what I have termed the crypticity of maternal effects. At the turn of the twentieth century, geneticists dubbed maternal effects "recondite," "unintelligible," and "capricious," and debated the limits of scientific methods for empirically testing claims asserting a link between intrauterine exposures and the later health and character of offspring. Despite radical transformations in the science of maternal intrauterine effects, questions of causal mechanism and disputes over what constitutes adequate empirical evidence for maternal effects remain at the center of concern today.

The struggle to specify causal pathways in today's science of intrauterine effects is humorously illustrated by a 2016 study that reported elevated voluntary exercise activity among the offspring of nine female mice submitted to fitness regimes during their pregnancy. The study gave rise to the *New York Times* headline "Does Exercise during Pregnancy Lead to Exercise-Loving Offspring?" "To some extent," claimed the *Times'* Gretchen Reynolds, "our will to work out may be influenced by a mother's exercise habits during pregnancy." But how to explain the finding? The article quoted a researcher speculating—in all seriousness—that "the mother's physical movements jiggle the womb slightly in ways that alter fetal brain development in parts of the brain devoted to motor control and behavior."[1]

In the 1990s, DOHaD founder David Barker's fetal programming hypothesis was derided by critics as "an inductionist's delight." "Example is piled on example, each somewhat consistent with the hypothesis but none seriously testing it," accused one.[2] A 2001 editorial in the *Lancet*, titled "An Overstretched Hypothesis?," charged Barker's exuberant productivity—61 papers over just a five-year period—with running far ahead of the "30 years plus" that would be required for a "truly prospective test."[3] Confident that the relationship between fetal exposures and longer-term health outcomes that they had observed must be causal, Barker's defenders argued that while "we do not yet understand the biological basis" of links between the intrauterine environment and heart disease in later life, "this lack of understanding does not deny the relation."[4]

The pursuit of causal explanations in fetal origins research today clusters around the possibility of identifying plausible physical mechanisms to explain the persistence of the effects of intrauterine exposures across lifetimes and generations. As chapter 7 demonstrated, central to this effort is detecting biochemical mechanisms—epigenetic markers chief among them—that might explain the persistence of the effects of those exposures over intra- and intergenerational time. In epigenetics, many fetal programming researchers see the prospect of a credible, material, manipulable agent connecting the maternal intrauterine environment to later offspring health outcomes.

But as I argue in this chapter, the arrival of more data and the identification of molecular markers that might constitute mechanisms for programming maternal effects at the level of the genome does not efface the central quandary of studies of the long-term implications of intrauterine experiences. To cohere molecular markers into a causal narrative also requires arguments for forms of causal inference that are more permissive than those conventionally prized within the biomedical sciences. This permissive approach to questions of causation gives scientists the room to ask big questions about potentially important developmental exposures that are nonetheless difficult to study. It also leaves more room for social assumptions and values to enter the chain of causal reasoning.

CRYPTIC EFFECTS AND THE SEARCH FOR A CAUSAL MECHANISM IN FETAL ORIGINS RESEARCH

Scientists are fundamentally interested in providing causal explanations. Isolating the contribution of causal factors to an outcome of interest is a centerpiece of good experimental design. It is also cardinal to larger public health goals of developing targeted interventions to improve health. When causality is well characterized, it opens the possibility of human manipulations to alter outcomes.

Today, unlike in earlier periods, maternal-fetal effects researchers can access expansive epidemiological data infrastructures, track a wide range of biomarkers indicating health status, and apply sophisticated methods for analyzing correlations within complex data sets. By coupling deeply dimensional longitudinal cohort studies that provide a rich mix of social, biomarker, and health data with the theory of epigenetic programming, fetal origins researchers hope to strengthen forms of causal inference in the field.[5]

The concept of epigenetic programming implies causal specificity in changes of levels of methylation in the genome.[6] It suggests that epigenetic changes convey information and transmit a spatiotemporal signal—that is, it implies specificity of location, function, and timing of methylation variation, and a reproducible, stable, and generalizable relationship between methylation variability at a particular site and an outcome. This specificity is a precondition for manipulability. Hypothetically, fetal programming scientists imply, researchers can discover the "program" and manipulate, control, and redirect it.

One way to situate epigenetic factors in a causal framework is to invoke the concept of a mechanism. A mechanism is an intermediate between a cause and an effect, providing information about how the effect comes about over time. Philosopher of science Carl Craver and colleagues argue in their account of mechanistic explanation in biology that a mechanism is best conceptualized as a process, a "series of activities of entities that bring about the finish or termination conditions in a regular way."[7] Mechanistic descriptions offer explanations that "render phenomena intelligible" and "show how possibly, how plausibly, or how actually things work."[8]

In DOHaD science, epigenetic programming of gene regulation is postulated as a mechanism physically linking prenatal exposures to later-life health outcomes. Epigenetic marks may be a component of the mechanism. Although the mechanism itself remains abstract, conceptualizing methylation changes as components or traces of a mechanism endorses the material reality and long-term stability of the maternal epigenetic imprint in causal relation to later offspring phenotype.

Classically, causes exhibit relations of dependence that are invariant, stable, and specific. Following the influential account of causation given by philosopher James Woodward, *invariance* means that under conditions of manipulation between the causal factor and specified dependent variables, a posited relationship will always obtain.[9] *Stability* implies that the causal relationship continues to hold under a wide range of background circumstances, including different times, places, and contexts. *Specificity* refers to the precision with which the cause is identified. Highly specific causes are switch-like, one-cause, one-effect relations.

A classic example is the paradigmatic "gene knockout" experiment, in which researchers show that destroying the function of a single gene leads, without exception, to the same functional deficit. In the case of epigenetics and maternal effects, consider the so-called agouti mouse, a genetically modified mouse strain with a retrovirus inserted upstream to the start site of *agouti* gene transcription for the trait of melanin pigment in hair color. The retrovirus locus in this mutant mouse is under tight epigenetic control via methylation. Dietary exposure during a critical window of intrauterine growth modifies methylation at this locus. In turn, the degree of methylation at that locus predicts mouse pup coat color (yellow, brown, or a range of mottling in between).[10]

Most cases of causality are not as clean-cut as the "gene knockout" or the case of the lab-engineered agouti mouse. But attributions of causality can be appropriate even in cases of complex multifactorial causality that are not switch-like single causes. This requires demonstrating—perhaps by degrees—the invariance, stability, and specificity of a causal relation against a backdrop of a particular context. Unfortunately, in the science of human intrauterine epigenetic programming, epigenetic findings are extremely unlikely to satisfy even these criteria. In the case of epigenetics, a truly causal claim minimally calls for an understanding of the interconnected cell regulatory and signaling networks in which variation at a particular CpG site participates.[11] But the issue is not merely the complexity of the biology, nor of the causal picture in human intrauterine exposures. As the detailed case studies in chapter 7 demonstrate, there is presently no evidence in humans that modest interindividual variations in methylation levels are induced by environmental factors during the intrauterine period, exhibit stability throughout the life course, and transmit specific information on which outcomes are causally dependent.

To be sure, some of this evidential deficit is because, practically speaking, these effects are challenging to study in human populations. The study streams analyzed in chapter 7—Project Ice Storm, Rachel Yehuda's Holocaust survivor studies, and the Dutch Hunger Winter studies— lack many elements necessary to establish causality, largely because of incomplete data. This includes well-matched and adequately powered controls; precise information about the timing and degree of the exposure; information about other early developmental exposures, including paternal, environmental, and early infancy; properly stored biological samples from the perinatal period such as cord blood or placental tissue; repeated biomarker assessments at multiple time points across the lifecourse; and biological samples from target tissues relevant to the hypothesis.

But even with the perfect study design and complete data of this sort, any causal story entertaining epigenetic variables will be highly tenuous due to the modest effect sizes, many co-factors, and difficulty of replication due to non-homology in observational constructs across varying human populations. Methylation variation seems particularly shaky ground on which to build a causal account of maternal intrauterine effects. All of the studies reviewed in chapter 7 solely report measures of average percentage methylation, the best-characterized and easiest-to-measure quantifiable epigenetic variable, surveyed in a subset of loci in the human genome. Recorded changes in levels of methylation are small-to-modest, ranging from 0.7 percent to 10 percent in compared study groups. At this time, we do not know whether detection of such small changes is reliable given the state of present technologies, whether such small changes can truly be stable over a long period of time, and whether such small changes could create a functional impact large enough to influence the distribution of diseases in a population.

As Bastiaan Heijmans, lead author with L. H. Lumey of several key studies examining epigenetic associations in the Dutch Famine cohort, writes, "Our basic understanding of the methylome is in its infancy. . . . Little is known about the actual scale and extent of between-individual variation in DNA methylation across the genome."[12] What is known affirms that "not all epigenetic variation is environmentally driven and some appears to be random in nature" and that "specific regions of the genome are characterized by stochastically variable methylation."[13] Furthermore, singling out methylation neglects other "layers of the epigenome," such as histone modification.[14] The technology used to detect and measure methylation levels is also limited, including fewer than 2 percent of CpG sites in the human genome and largely "focused on promoters and CpG islands that may not be the most relevant."[15] That is, there are questions about the ability of this technology to reliably detect moderate-to-small effects on methylation levels, which characterize human findings. Combining these concerns with the small sample sizes, lack of well-defined exposures, weak longitudinal designs, and the use of blood tissues that may not concord with methylation levels in tissues relevant to the phenotype, Heijmans concludes that "we may be trying to detect inherently small effect sizes using suboptimal methods and sample cohorts."[16] All of these considerations—known and unknown features of the biology of epigenetics; technological, statistical, practical, and ethical constraints; the many environmental confounders of early human development; and persistently small effect sizes—help constitute what I have characterized as the crypticity of maternal-fetal epigenetic effects.

Findings of epigenetic modifications are frequently portrayed as strengthening the plausibility of a causal association between gestational exposures and later outcomes. But, as I have argued here, the facticity of methylation as a quantifiable, measurable molecular adduct with the precise chemical structure of $-CH_3$ does not establish it as a causal agent. In reality, even the most cited examples of promising work in human maternal-fetal epigenetic programming science cannot support causal claims. This is because the effects are cryptic: that is, effect sizes in health outcomes in this arena are small; the causal chain between exposure, epigenetic marker, and outcome is temporally attenuated; and we know little about the function, stability, and reversibility of methylation changes at particular sites in the genome. In short, identification of epigenetic changes correlated with observational associations between a fetal exposure and later outcome does not confirm causation, nor does it reduce the crypticity of that original association.

How, then, do fetal programming scientists persist in asserting strong causal claims in the field of fetal programming research, despite the unresolved crypticity of both epigenetic facts and the maternal effects they purportedly explain? In what follows, I argue that social gender norms about the unique maternal agency and responsibility for fetal outcomes bestow unusual latitude to DOHaD scientists to sustain the conviction that associations between intrauterine events and later outcomes must be causal.

SOCIAL VALUES, GENDER BELIEFS, AND CAUSAL REASONING

It is widely accepted that assignment of causality, both mundane and scientific, can be influenced by social context, subjective human judgment, and appeal to shared beliefs and values. On the first order, what is dubbed a cause is often linked to human practical interests in manipulation, control, and intervention. As Woodward argues, we call things causes, and focus on those causal factors, that appear to us to be "potentially exploitable for purposes of manipulation."[17] Additionally, causal claims, as Woodward puts it, have a "contrastive structure," such that "to causally explain an outcome is always to explain why it rather than some alternative occurred."[18] Human cognition and social norms can influence our intuitions about what *should have happened* and our evaluations of what is normal or optimal. Finally, judgments of causal relevance can also be understood as social in the sense that our socially learned, culturally inflected expectations about chronological regularities in the world influence "what we take to be a serious possibility."[19] As a result, Peter Machamer and colleagues argue, "What is taken to be intelligible (and the different ways of making things intelligible) changes over time . . . intelligibility is historically constituted and disciplinarily relative."[20]

It is no coincidence that in her classic article on the social dimensions of causal assessments, "Context Effects in Judgments of Causation," social psychologist Ann McGill used the gendered scenario of attributing causality to a teenager's pregnancy to flesh out these very ideas. She begins with the following question: What factor—sexual intercourse, lack of birth control, or the fertility of both the man and woman—caused a teenager's pregnancy? McGill goes on to unpack, using experimental findings, how social and other context influences causality attribution. The fertility doctor, the teen, the parents of the young woman who is pregnant, or the sex educator each locate causality differently. What people pick out as causal factors, McGill demonstrates, are those details that deviate from their expectations, informed by their social location. Social norms can "govern comparative judgments and designate experiences as 'surprising,'" and this can lead to different readings of causality in the same episode, McGill concludes.[21]

Ascriptions of causal responsibility can also vary depending on how the scenario is framed.[22] Cognitive psychologist Tania Lombrozo recently examined this phenomenon in a headline-making episode in which police arrested an African American mother in South Carolina for leaving her child alone for an hour at a playground while she went to work. Lombrozo pointed out that harsh attributions of responsibility to the mother alone, in this case, reflected race- and class-based assumptions about the range of alternatives available to the woman. These judgments failed, for instance, to extend charity to the mother's judgment about the safety of the playground and maturity of the child, to ask about the role of the father or other co-parents, or to consider the way that inconsistent shift worker schedules, lack of quality low-cost daycare services, and work requirements for individuals receiving government assistance might create an impossible scenario for the mother. As Lombrozo argued, "When dealing with complex causal processes and the assignment of causal responsibility ('it's the mother!'), values can affect the conclusions we draw from science in an especially pernicious way. That's because we *think* of causal claims as simple descriptive facts about the world—as value-free. But a growing body of empirical work shows they're not. In fact, the way we make causal claims depends a lot on how things *normally* happen and on how we think they *should* happen."[23] Implicit assumptions and expectations regarding women's obligations to optimally nurture offspring may lead us to overattribute causality to mothers "even when other causal factors were also at work and even when an individual mother may not be the most appropriate locus for intervention."[24]

Social assumptions can especially influence ideas about causal relevance when the behavior at issue has a powerfully negative valence. Social epidemiologist and maternal-infant public health advocate Lawrence Wallack, collaborating with cognitive psychologists, has shown that in public health discourse, the mere suggestion of "harm of a mother to a child"—or fetus—is an especially overpowering social and psychological frame. The image of a mother harming an infant immediately shuts down social imaginaries about the larger causal picture—narrowing the perception of avenues for intervention.[25] As Wallack and colleagues write, "Asking, 'What would a woman do today if she wanted to help her baby avoid chronic disease?' is very different from, and much more limiting than asking, 'What would our society do and provide if we wanted to be the healthiest place to be born?'"[26]

In sum, powerful and entrenched assumptions about maternal responsibility can structure our causal explanations and limit the conceived space of causal alternatives. Of course, these considerations are not limited to social science scenarios. Biomedical researchers frequently make judgments of causal relevance among a wide variety of interacting causal factors that together form a chain or causal structure leading to the outcome of interest. How does this play out in the science of fetal programming and developmental origins?

"IT'S THE MOTHER!"

A typical visual representation of the epigenetic fetal programming hypothesis is a tangle of arrows. The arrows represent theoretical pathways connecting maternal and fetal bodies to multiple ecologies and forms of exposure along a developmental trajectory. Despite what appears to be a complexity-affirming visual convention, these representations also reveal that causal attributions in DOHaD explanations accumulate principally around the mother's body and behaviors during pregnancy. Reflecting those attributions is one consistent trope of DOHaD arrow diagrams: a central, prominent profile of a transparent, nude, headless, human female with an extremely pregnant abdomen containing an infant-sized fetus. Arrows flow into the abdomen, representing preconceptional and in-utero exposures, and then out of the abdomen, leading to the programming of the epigenome and later-life risks.

In one such diagram, displayed at a National Academies of Science meeting on childhood obesity that I attended in Washington, DC, in 2015, all arrows eventually lead through the maternal body and land at a cartoon of a little ambulance, lights blaring, rushing her afflicted child to the emergency room (fig. 8.1).[27] Although the diagram includes the possibility of multiple environmental exposures, the pregnant female body and its fetus is the vivid center of the picture—the only human image among the boxes, arrows, and molecular symbols positioned as the agential conduit and point of intervention for preventing that emergency room visit. Another presentation at the same conference dispensed with the arrow salad entirely, abridging this diagrammatic convention to portray a single large, bold arrow flowing from a photograph of a headless pregnant woman zealously enjoying a burger and fries to an image of an obese youngster.[28] The diagram portrays the pregnant woman's behavior as the direct cause of the same evidently deplorable behavior in her offspring.

This focus on maternal intrauterine effects in DOHaD is reinforced at every stage of research. A 2018 analysis by epidemiologists Gemma Sharp, Debbie Lawlor, and myself published in the journal *Social Science and Medicine* found that, in the field of DOHaD, 20 times more papers treating maternal effects have been published than those treating paternal ones. Leading prospective cohort studies beginning in the fetal period, such as Project Viva in the United States and Generation R in

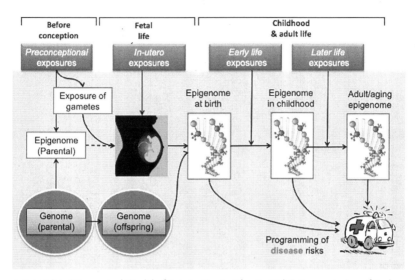

FIGURE 8.1 "Conceptual Model of Epigenetics in Obesity Risk." By permission of Andrea Baccarelli.

the Netherlands, collect only the barest if any information about fathers, via the mother, and report results of associations only between maternal exposures and offspring outcomes.[29] Even major research studies that make a concerted effort to include fathers, like the UK-based Born in Bradford study, collect far greater and better-quality data on mothers than on fathers. Begun in 2007, Born in Bradford recruited 12,453 mothers and 3,356 partners.[30] Mothers completed 22 questionnaires as well as providing multiple clinical measurements and biological samples, whereas partners completed one questionnaire around the partner's pregnancy or the birth of the child and provided one biological sample.[31] Researchers defend these differences by claiming that fathers are harder to find and keep enrolled in studies. If that were the only factor in their understudy, we would expect animal studies, in which fathers can easily be sampled, to include paternal factors at much higher rates. But, as our team showed in a subsequent 2019 analysis of articles in the *Journal of the Developmental Origins of Health and Disease*, 90 percent of animal studies considered only maternal factors.[32]

Part of the challenge is that pressures to publish and obtain funding keep researchers coming back to questions for which there are existing data sets and deep publication lineages, rather than venturing into riskier new areas, and these pressures also encourage the publication of positive results rather than null or negative ones. This in turn influences the data that the field continues to gather and the kinds of research questions it considers to be important and tractable, thereby inflating the importance of maternal pregnancy effects as determinants of offspring outcomes in a reinforcing cycle (fig. 8.2).[33] This overwhelming focus on maternal pregnancy effects—in isolation from paternal, postnatal, and wider social and environmental factors in development—is based on implicit, often unquestioned starting assumptions about the causal primacy of maternal effects, relative to other possible factors, in producing later outcomes. These assumptions are entrenched at every stage of research design, analysis, publication, and public translation.[34]

Fetal programming theories endow women's reproductive bodies and behaviors with profound explanatory importance; at the same time, paternal bodies and wider communities recede into the background. A clear example of the systematic bias that this introduces into the science

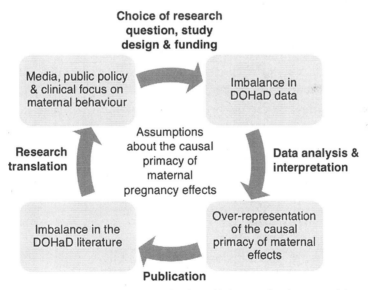

FIGURE 8.2 Assumptions that the health, lifestyle, and behaviors of mothers around the time of pregnancy have the largest causal influence on their children's health and risk of disease drive DOHaD research at all stages, from study design to research translation. From Sharp, Lawlor, and Richardson, "It's the Mother! How Assumptions about the Causal Primacy of Maternal Effects Influence Research on the Developmental Origins of Health and Disease."

itself is given by the frequency with which DOHaD studies test maternal pregnancy effects in isolation from other factors, whether in humans or animals. Of 325 eligible articles published between 2010 and 2018 in the *Journal of the Developmental Origins of Health and Disease*, 84 percent describe studies of maternal exposures, and 78 percent of those treated maternal exposures in isolation.[35] By comparison, only one study in the entire data set considers paternal exposures in isolation.

Observations of associations between maternal intrauterine exposures and offspring outcomes in isolation are insufficient to conclude that the association is plausibly causal, since confounding genetic, lifestyle, socio-economic, environmental, paternal, postnatal, or other factors may also correlate with the exposure and better account for the findings. When researchers include other variables and apply methods designed for strengthening causal inference in epidemiology, health outcomes once traced exclusively to maternal intrauterine effects are often challenged. For instance, a study of 10,000 children that associated poor maternal health during pregnancy with childhood intelligence found that when other factors were added, the most important predictor was father's social class.[36] Studies have frequently claimed that maternal age, drinking, or smoking in pregnancy is linked to socioemotional and cognitive development, but a study that included postnatal factors demonstrated that exposure to long-term poverty had a greater effect on these outcomes than any maternal pregnancy factor.[37]

Including fathers as negative controls proves to be a powerful method for assessing whether maternal pregnancy effects are causal. A research group that previously found maternal smoking during pregnancy to be associated with ADHD in children discovered, when fathers were included in the study design, similar effect estimates for maternal and paternal smoking around the time of pregnancy, suggesting that shared family genetic or lifestyle characteristics may better explain the association.[38] Perhaps most striking are challenges to the theory that maternal obesity during pregnancy leads to higher rates of obesity in their children—by any measure, a core, oft-repeated claim of DOHaD research.[39] Several efforts to test this association that include paternal as well as maternal BMI have shown that paternal and other confounding factors better explain later offspring risk of being overweight than does maternal BMI at the start of pregnancy.[40]

"THE WOMB MAY BE MORE IMPORTANT THAN THE HOME"

DOHaD's conceptualization of the maternally provided intrauterine environment as a primary cause of adult outcomes, and of pregnancy as the most important period for programming epigenetic set-points in the genome, conveys a drastically limited picture of influences on development, effacing structural, social, and environmental determinants of health across the life course. "The womb may be more important than the home," David Barker famously quipped.[41] But vast evidence suggests that it is quite the opposite.

The home, the community, and the many-textured aspects of the social world are likely far more important than prenatal variables in determining the outcomes of interest to DOHaD researchers. As sociologist Dorothy Roberts concisely puts it, "The best predictor of health is an individual's position in the social hierarchy."[42] Intrauterine indicators, whether birth weights or methylation signatures, are confounded by these social factors. As a result of that confounding, associations between those indicators and later life outcomes may have no causal connection; the two may instead be the result of an individual's social trajectory.

"Virtually every aspect of reproduction and parenting is culturally mediated, socially patterned, and structurally informed and constrained," writes University of Michigan social epidemiologist Arline Geronimus, "while simultaneously having potential biological impacts on growth, development, and health."[43] An excellent example of this can be seen in another highly contested area of scientific literature on parenting—breastfeeding.[44] Breastfed infants are said to have lower rates of obesity, diabetes, and hypertension, but these findings are confounded by the fact that breastfeeding parents are different than those who choose other methods. Notes Geronimus, they "are more highly educated, economically better off, disproportionately White, have sufficient job flexibility to breastfeed, have a cultural commitment to and social support for breastfeeding, and have sufficient nutrition and low stress levels to produce the necessary milk supply."[45] DOHaD researchers' failure to rigorously attend to this sort of confounding in the case of intrauterine exposures leads to "estimated associations between early-life health indicators and later-life outcomes" that "exaggerate the causal effects of early-life exposures," argues Geronimus.[46]

Even if intrauterine factors are causal in adult diseases—an open question—it is almost certain that home is more important than the womb. That is, postnatal factors and later life cumulative exposures are likely far more important than small variations in the intrauterine environment. Extensive evidence supports the view that distributions of health risks across populations reflect negative health impacts accumulated across a lifetime. Non-European immigrants to the United States, for instance, generally show better health than ethnically related United States citizens. But the health of later generations deteriorates so that it resembles those of ethnic minorities in the United States. These changes, Geronimus observes, reflect "social realities in the United States that have measurable effects on the health of those who developed in (or whose mothers developed in) less hostile uterine environments, and also on those who arrived as adults, already older than proposed critical developmental windows."[47]

The understanding of the social dimensions of health inequalities embraced by fetal programming science deemphasizes the role of structural inequalities throughout youth and adulthood, focusing instead on molecular determinants set before birth. As Roberts puts it, in this research, "scientists convert social inequality into a series of deleterious environmental exposures" and "developmental deficiencies."[48] In maternal-fetal epigenetic programming science, that environment is a human woman, and the deficiencies a direct result of her cues to the fetus during prenatal development. "The social order," writes Roberts, "already begins to fade from the model as attention is focused on biochemical responses within the body rather than on the inequitable social structures the body is responding to."[49] Conceptualizing these biochemical responses as programmed, permanent, and even heritable, the result is a picture of fetuses inflicted with "self-perpetuating biological deficits" invoked to "explain what prevents oppressed people from escaping their disadvantaged position in an unjust social order."[50] The outcome of this conceptual scheme, writes Roberts, is not merely failure to factor in broader social factors but active "obfuscate[ion]" of the "existence of ongoing deficits."[51]

Attributing poor outcomes to "self-sustaining biological deficits" directs attention toward the social cost of such deficits and focuses efforts on regulating, fixing, and intervening on mothers rather than on social structures. As a result, the policy solutions that emerge from DOHaD often focus on maternal behaviors without appreciation for social realities, similar to offering unrealistic advice that tells poor mothers to breastfeed without considering social context. Destined to fail because, as Geronimus points out, no "bubble exists for pregnant or lactating mothers to inhabit apart from the stressors they cope with throughout their lives," such DOHaD messages may have unintended negative consequences, reinforcing "the popular framing of the problems of the poor as self-inflicted."[52]

IMPROVING MATERNAL AND INFANT HEALTH OUTCOMES: THE BIG PICTURE

The stakes of our judgments of causality in fetal origins science are high. Findings in maternal intrauterine effects science are rapidly translated for public consumption, in part because any implication of maternal-fetal harm packs an unmatched emotional and political intensity.[53] Add to this the intrigue of epigenetics—often promoted as a radical new way of understanding human agency over our biological destiny—and you have instant pop science catnip. From the *Time* magazine cover story "Why DNA Isn't Your Destiny" and *NOVA* television special "The Ghost in Your Genes," to books with titles such as *The Epigenetics Revolution: How Modern Biology Is Rewriting Our Understanding of Genetics, Disease and Inheritance*, epigenetic science has attracted tremendous popular attention over the past decade, with maternal-fetal epigenetic programming as Exhibit A.[54]

We live in a world with massive inequalities that drive disparities in maternal and infant health.[55] In the big picture, the most significant predictors of poor outcomes for fetuses and newborn infants are high rates of preterm birth and very low birth weights among disadvantaged populations. In the United States, these differences are stark—African Americans have 2.3 times the rate of fetal mortality of white women, and a 59 percent higher rate of preterm births compared to whites.[56] Globally, 98.5 percent of neonatal deaths occur in developing countries. The rates are highest where prenatal care is minimal and births are not attended by skilled personnel.[57]

How we understand and evaluate causality in epigenetics and fetal origins science matters. If we set out with the shared goal of improving the lives of women and infants, what interventions are most likely to move us toward that goal? Most advice to individual women is only moderately-to-poorly supported by strong evidence, and the effect sizes are small.[58] But there is one intervention that approaches a silver bullet: Experts in neonatal mortality universally recognize access (or lack of access) to health clinics with high quality care as the single most important driver of neonatal health.[59] This claim is supported by historical evidence in the United States, where free clinics in poor, African American communities have dramatically improved maternal and infant health outcomes. This evidence points to the importance of structural racism and poverty, rather than women's behaviors, as the overwhelming causes of poor gestational outcomes.[60] Not only do community health services greatly reduce fetal and infant mortality, but they also improve maternal outcomes. Trained birth workers, including doulas, who serve as birth companions, coaches, and advocates, sharply reduce the rates of C-sections and maternal mortality.[61]

During a reception at the annual Society for Maternal-Fetal Medicine meeting in San Francisco in 2013, I struck up a conversation with a physician-researcher, who claimed to have held one of the first fellowships in the maternal-fetal medicine specialty (which formed in the 1970s). I asked him how research in the field had evolved over the decades. His answer surprised me—what's more interesting, he said, is how little has changed. Other than folic acid supplementation, he claimed, decades of research have yielded clinicians few interventions that empirically improve infant outcomes. Anthropologist and leading DOHaD researcher Christopher Kuzawa sounded a similar note at a 2016 meeting that I attended on the developmental origins of health and disease, hosted by the Human Capital and Economic Opportunity Global Working Group at the University of Chicago. Kuzawa observed in his opening remarks that after twenty years of productive research, DOHaD studies had not yet translated to a single clinical or public health intervention that could be scientifically shown to improve life chances.[62]

Empirical attempts to assess whether improvements in fetal health lead to better life outcomes have come to similar conclusions. This is particularly clear in the case of the vast body of research correlating birth weights with adult outcomes. Health economist Eric Schneider has shown that mean birth weight and distribution has remained stagnant, despite massive gains in health status. Over the past 150 years, infant mortality has plummeted, and adult male stature has increased by as much as 2.3 standard deviations. At the same time, we saw higher rates of twins, first births, and surviving very low-birth-weight infants—all of which should have lowered average birth weights. Yet average birth weight has stayed the same since the late nineteenth century in North America and Northern and Western Europe. Even the decline in smoking prevalence—also associated with low birth weight—registers not at all in the historical birth weight record. On the one hand, perhaps this offers further evidence that birth weight is an exceptionally poor measure of improvements in health in utero (see chapter 6). On the other, if, despite large gains in human health, medical technology, and standard of living, birth weight has remained stagnant, Schneider's findings may suggest that the posited relationship between fetal growth and later health and life outcomes is at best an exceptionally cryptic one.

Among the strongest causal claims asserted in the fetal programming literature is David Barker's argument that improving fetal growth could prevent adult heart disease. Epidemiologists Diana Kuh and Yoav Ben-Shlomo applied the effect sizes found in Barker's studies to query how much additional fetal weight gain would be required to make a dent in heart disease rates, assuming Barker's findings are valid.[63] They found that all births would need to weigh between 9 *and 9.5 pounds*—up to two pounds greater than the current average—to match the magnitude of impact of already highly efficacious interventions (e.g., smoking cessation or lowering cholesterol). The same was true with systolic blood pressure. Based on Barker's numbers, Kuh and Ben-Shlomo found that it would require a greater than two-pound increase in average birth weights just to yield a very modest 1–3 point decrease in adult systolic blood pressure (falling far short of the 15-point reduction in systolic blood pressure that would be necessary to lower heart disease risk by 50 percent in an adult).

Not only are such increases unattainable—an increase of even a quarter of a pound is rare among studies that have attempted such interventions through nutritional supplementation or counseling—but they are undesirable. Raising every birth to 9–9.5 pounds, Kuh and Ben-Shlomo point out, would bring health trade-offs. For example, larger babies would increase the use of risky interventions (such as C-sections, forceps, and oxytocin) in labor, increase rates of shoulder dystocia during birth, and require greater weight gain during pregnancy, significantly raising post-partum risk of obesity for pregnant people. Moreover, while lower birth weight may be associated with higher cardiovascular disease risk in adults (though whether that association is causal is still in question), higher birth weight is equally strongly associated with ovarian, prostate, breast, and other cancers in adulthood. Should we reduce cardiovascular disease at the risk of increasing risk of these diseases?

Because interventions to increase fetal growth offer marginal health benefits while increasing unintended adverse health effects, Kuh and Ben-Shlomo concluded that "intervening to improve fetal growth based solely on the programming hypothesis is not justified."[64] They also note an additional potential harm. Directing interventions toward optimizing outcomes for these births, Kuh and Ben-Shlomo warn, "could detract from the . . . focus on preventing preterm, very low-birth-weight infants by shifting the emphasis to normal weight infants born at term."[65] This point is worth underscoring. Fetal programming research focuses on small variations in measures of fetal development in full-term, normal-weight infants. As a public health priority, a focus on increasing birth weight among normal-weight infants born at term redirects resources and the locus of concern from a much higher priority for community health—preventing preterm births.

While diffuse claims about cryptic maternal-fetal influences have created little by way of effective interventions to improve health outcomes, they have contributed to parents' fear and anxiety, and to stigma and bias against particular pregnant people. Take, for instance, investigations of the relationship between the maternally provided pregnancy environment and obesity, often represented as a global public health epidemic carrying high social and economic costs with few effective interventions.[66] DOHaD scientists hypothesize that maternal nutrition and metabolic status leave an epigenetic imprint on the growing fetus that contributes to adult disorders.[67] Driven by this hypothesis, DOHaD studies have examined every possible correlation between maternal obesity during pregnancy and children's health and life outcomes, including in hot button areas such as autism, ADHD, and brain size.[68]

News coverage of this research favors positive findings that affirm the harms of maternal obesity. Exposure to obesity or a high fat diet during the womb, relayed one typical news article that caught my eye when I was pregnant, "can have permanent, pervasive effects on the child's health."[69]

"You Are What Your Mum Ate," read a 2012 ABC News headline,[70] while a 2014 *Science* news blurb on a recent finding on fetal metabolic programming was headlined "The Nutritional Sins of the Mother."[71] A 2012 paper showing that rats whose mothers ate a high fat diet during pregnancy had a higher chance of cancer compared to those who did not led to the headline, "Why you should worry about grandma's eating habits," and admonitions to pregnant women to "Think twice about that bag of potato chips."[72] (News coverage did not explain that the rats were bred for high cancer rates, nor that cancer rates actually went down among descendants of those born of high-fat-diet pregnancies, relative to controls.)

Sociologist George Parker has interviewed obese pregnant women at the receiving end of DOHaD-related claims about the harmful imprint of their obese bodies on their future children and descendants. She found that "the intense problematization of their weight left participants deeply worried about the potential harm posed to their babies by their fatness, leading some to question whether they should ever have become pregnant in the first place and others to grapple with the idea that they had failed at being a mum before they had even begun."[73] Parker reported that exposure to such claims increased health risks for the women, leading to a gruffer and lower standard of care for obese women, encouraging obese women to distrust and disengage from medical advice, increasing rates of depression and poor body image, and excluding them "from low-risk care options that promote low interventionist birth."[74] Ultimately, Parker's interviewees reported "feeling as though they have hurt their babies and burdened society, resulting in a deeply negative affective space . . . in their transition to parenthood."[75] Parker laments the role that fetal programming claims have played in "entangling pregnant people in a politics of preemptive action to secure their children's future health, all the while denying the sociopolitical, economic and cultural realities of women's lives that constrain their ability to do so."[76] In the case of claims about obesity, we can add to these concerns a worry about replicating harmful stereotypes and misconceptions that contribute to stigma about fat children, which in itself can harm their mental and physical health and imperil their safety. The notion that a fat child is something to mourn and to avoid at all costs is implied by scientific claims common in the fetal programming and DOHaD literature.

The implications of these discourses for the embodied experience of pregnancy should not be underestimated. Presently, more than one-third of adult women of reproductive age are categorized as obese.[77] What might the world look like if we helped people of all kinds, with all kinds of bodies and health statuses, have healthy pregnancies, instead of stigmatizing some as inadequate vessels threatening their children, society, and future generations? Parker's interview subjects yearned to "tell their stories of pregnancy and birth in more affirming and helpful ways," for the freedom to experience, through pregnancy, "new knowledge of their bodies' strength and capacity," to be "treated with dignity and respect," for "replacing approaches to pregnancy health that produce blame and shame with those that promote care and celebration," and for more emphasis on the "shared responsibility and access to the material conditions need to secure healthful pregnancies."[78] Parker argues that, in the end, such an approach is far more likely to help pregnant people engage with health advice than one that presents certain bodies as hurting their fetuses and harming society.[79]

This is not to deny that empirically driven, causally sophisticated investigations into how maternal-fetal effects shape health outcomes may someday produce information that improves fetal outcomes in human populations. The existence of bigger nuts to crack in the arena of maternal and infant health does not make the sorts of cryptic phenomena found in DOHaD research irrelevant. There is inherent value in this knowledge, even if the effects under study are limited to narrow contexts and are relatively small in magnitude, with causality in question.

Still, pregnant people need information to evaluate prenatal risks and make choices about what is best in the context of their lives. Risk is everywhere. Parents must work, drive, exercise, eat, and endure stresses ranging from the everyday to the extraordinary. Parents of every class and race are equally likely to be concerned for their children and wish for their well-being. What is needed is knowledge that empowers. Maternal-fetal effects science needs to give an accurate picture of the degree of certainty in estimates of risk, and parents need the ability to weigh these risks in relation to one another, and in relation to other goals and values in life. The starting stance cannot be that any risk is unacceptable, nor that all exposures are within a parent's control.

If scientists do not attend to these issues, fetal programming science will function as another stigmatizing, determinist discourse that threatens women's reproductive autonomy and contributes to stigma and moral panic about the mothering practices of poor women, women of color, and non-normative mothers of all kinds.[80] One could imagine a future in which prenatal and preconceptional exposures of all sorts—maternal, paternal, and other relevant environments—are explored in tandem and weighed in context, without a prior construct driving overwhelming focus on the mother.[81] But to get there, the field of fetal programming science and DOHaD must rebuke the frame of maternal-fetal harm and commit to ending the practice of decontextualized causal risk claims.[82]

All of us need deep humility about the crypticity of the causes and effects in maternal effects science, and we need frank talk about the potential for fetal epigenetic programming and DOHaD research to contribute to an imbalanced focus on fetal exposures compared to other risk factors. It is a fallacy to accord higher causal plausibility and primacy to a factor simply because it can be rendered as a quantifiable level of methylation on a gene chip and described in biochemical terms. In the end, we must keep in sight the bigger picture of the largest, most pressing priorities for both mother and offspring health, for which the evidence is already in: equitable distribution of prenatal, birth, and postnatal care; access to nutrition, essential medicines, and education; stable income; housing; and freedom from violence.

CONCLUSION

My approach to the variegated history of maternal effects science in this book has placed the social processes of scientific knowledge production and validation at the center of inquiry. Examining selected scientific claims in this field and the debates around them, I have sought to illuminate how judgments of validity of inference and evidential relevance reflect gendered contextual background assumptions about what is scientifically knowable, possible, and plausible.

In my reading, the central drama of the science of maternal effects over the past century is a shift from a view of maternal environmental effects as largely scientifically unknowable, to a view of them as observations about which science can make causal inferences. Previously considered small, relatively unimportant compared to other influences, and in any case largely inaccessible to scientific methods for ascertaining causality, today scientists take a far more expansive and permissive view of maternal effects, as pervasive and powerful drivers of life outcomes that are—at least in principle—testable using scientific methods of causal inference. That is, over the past century, the scientific community shifted its stance regarding which sorts of claims are permissible and upon which science can pronounce. To use Weismann's formulation, maternal effects have gained both the "right to be looked upon as scientific facts" and the right to be formulated "as scientific questions."[83]

As I have argued, this shift is not explained solely by the accumulation of new data, technologies, and causal reasoning methods in recent decades. After all, the result of these efforts remains cryptic effects that do not substantiate a causal link between maternal intrauterine environment and later outcomes and have not led to interventions that improve health outcomes for parents or offspring. The foundation for this dramatic shift is a theoretical framework, advanced by scientists, funders, and allies in public health and related fields beginning in the mid-twentieth century, articulating early development as the critical playing field for the environmental determination of human variation in life outcomes. A vision of maternal effects as an ameliorative biosocial science further helped to open a more permissive space for projections about the long reach of the intrauterine environment. Within this framework, the maternal intrauterine environment is positioned as a vector for adversities that leave a permanent negative impact on the vulnerable fetus, magnifying the intransigence of social inequalities. This reasoning combines with powerful social assumptions about tolerable levels of fetal risk and the acceptability of restrictions on women's reproductive autonomy to contribute to greater permissiveness toward claims-making about cryptic effects at the maternal-fetal interface.

Given the crypticity of human maternal effects and the central role of human beliefs and values in the evaluation of the strength, plausibility, and promise of scientific theories and claims in maternal effects science, it will not be enough just to go where the evidence takes us. Resolving the question of the extent and importance of maternal intrauterine effects in human populations is not merely a matter of persisting through a long period of empirical uncertainty. Because our assumptions and social values affect our understanding of available data, the way forward entails opening the science to wide critical examination—with the acknowledgment that judgments informing causal inferences are never outside of reproductive politics in the context of ongoing racial, gender, and class inequalities.

CHAPTER 9

Epilogue: Gender and Heredity in the Postgenomic Moment

The 2001 terrorist attack on the World Trade Center in New York City caused the deaths of nearly 3,000 people, widespread economic perturbations, a new tide of anti-Muslim sentiment in the United States, and the widening of US military interventions in the Middle East. It was also a signal event that initiated North Americans into a post-terrorism subjectivity characterized by a pervasive sense of risk and uncertainty about the future. Immediately after the attacks, the American College of Obstetrics and Gynecology issued a warning that "pregnant women might be particularly vulnerable to the event's impact."[1] In the decade that followed, studies appeared in leading journals of public health and human reproduction reporting that infants in utero at the time of the 9/11 attacks bore the mark of the disaster, with implications for future generations to come.

One study discerned "significantly increased risks for low birth weight among Arabic named infants" in California.[2] Another found distinct birth weight and developmental outcomes among children of lower Manhattan mothers who were pregnant at the time of the attacks and who showed symptoms of post-traumatic stress disorder.[3] Surveying these findings, a group of physician-scientists studying the public health impact of 9/11 projected that "long-term follow-up of the infants exposed in utero . . . will be required to characterize and manage the long-term and delayed consequences of the disaster and thus to chart the full magnitude of the health impact of the attacks on the [World Trade Center]."[4]

One hundred years ago, in the first decades of the twentieth century, the claim that communal anxiety provoked by a terrorist attack could impact the development and future fates of growing fetuses would have been forcefully rejected by most biological scientists and medical doctors. They relegated notions that maternal experiences can imprint on the fetus to the realm of old wives' tales and premodern mysticism, and to a minority, perceived as superstitious and pseudoscientific, still clinging to Lamarck's theories of social and environmental inheritance. But today, the science of maternal-fetal effects has become one of the fields of speculation through which we understand and process the risk and uncertainty of our world, and research on fetal epigenetic programming is at the leading edge of biological investigation across the fields of genomics, public health, psychiatry, and reproductive medicine.[5]

In contrast to prenatal culture theorists whose efforts focused on advice to middle-class parents for the optimization of their families, today's maternal intrauterine effects researchers see their work in a biosocial frame that is explicitly tied to social justice aims of directing resources to the disadvantaged and communicating the harms of social inequality, racism, war, famine, and abuse.[6] "We now know that the quality of life directly affects how our genes operate," states public health professor Larry Wallack in a 2014 film titled *Epigenetics and Equity*. "What epigenetics tells us is that these diseases rather than starting later in life actually occur at the earliest stages of development and probably even in the previous generation. People are being born into this world at fundamental disadvantage. . . . This is not just a health problem. It is a broader social equity issue."[7]

Dorothy Roberts characterizes epigenetics as a "new biosocial science," explicitly positioned as a progressive, democratic science that openly aims to help policy makers "confront the reality of social inequality."[8] Committed to the hypothesis that the social gets "under the skin"—what epidemiologist Nancy Krieger has memorably dubbed "embodied inequality"[9]—new biosocial scientists present molecular methods such as epigenetics as powerful tools for illuminating "biological consequences of social inequality at the molecular level" and discovering the "biological pathways through which social inequality gets embodied."[10]

Although, as this book has shown, the search for a progressive biosocial science at the maternal-fetal interface has earlier roots, a landmark in the present-day reassertion of this aim was sociologist Dalton Conley's 2003 book, *The Starting Gate: Birth Weight and Life Chances*. Conley outlined a vision of a progressive, democratic biosocial science focused on the well-being of children and mothers and using scientific knowledge to illuminate and redress the harms of poverty and racism.[11] Aiming to analyze "how society, biology, and genetics interact with one another across generations," Conley and coauthors set out to redress sociologists' neglect of "biological and genetic factors in social life for fear of being called biological determinists."[12] Conley focused on birth weight, which he characterized as a "rich index . . . that may be exploited for intergenerational sociobiological analysis," to launch a compelling case for the profound ways in which the social shapes our biologically lived experiences, with implications for individuals' future achievements and those of later generations.[13]

Today, Conley's vision is widely shared. A growing body of human epigenetics findings report correlations between adverse early developmental events, biological variables such as methylation levels, and health outcomes such as depression, mental illness, and cardiovascular disease.[14] Responding to this flowering of biosocial science, in 2016, the New York–based Russell Sage Foundation introduced a new initiative called "Integrating Biology and Social Science Knowledge." The foundation describes the biosocial turn in its call for applications as "a paradigm shift in the life sciences, spurred by the realization that many biological processes, rather than being fixed, immutable mechanisms that consign people to particular life outcomes, are instead fluid, dynamic responses to features of the social and physical environments humans inhabit." This shift, the foundation writes, has "led researchers to launch interdisciplinary studies that seek to integrate approaches from the social and biological sciences, recognizing the potential for a deeper understanding of how social inequalities are initiated, maintained, and transmitted."[15] In the UK, the British Economic and Social Research

Council, jointly with the Biotechnology and Biological Sciences Research Council, has also announced a dedicated biosocial research funding scheme. The council defines biosocial research as "concerned with the dynamic interplays between biology, experiences and behaviours over the life course." Like the Sage Foundation, the council emphasizes the integration of biological, medical, and social science approaches, all oriented toward a horizon of public health and policy applications.[16] Studying how "adverse experiences, for example short and long-term consequences of poverty or parental divorce, or 'scarring' and health effects of spells of unemployment . . . 'get under the skin' to alter biology at one or more levels," wrote one scientist on the initiative's blog *Society Now*, "will enable improved targeting of interventions . . . and lead to more precise ways of reversing these consequences."[17]

Human maternal-fetal epigenetic programming science coheres cryptic findings into bold biosocial theories, which locate causality and agency in the intrauterine period. Epigenetic explanations collapse scales of biological materiality from the molecular to the macrosomic, levels of biological and social analysis from metabolism to wartime trauma, and temporalities of individual memory, generational inheritance, and cultural history into highly specific, quantitative claims about levels of DNA methylation. The result is a collection of compelling biosocial narratives charged with urgency by the specter of fetal harm and with promise by researchers' pronouncements that a forthcoming applied science of fetal programming can redress it.

Visions of epigenetics as a transformative framework for biology and the social sciences are nowhere more powerfully developed than in the pronouncements of McGill University epigeneticists Michael Meaney and Moshe Szyf. Meaney writes that epigenetics is "likely to have profound consequences when you start to talk about how the structure of society influences cognitive development. We're beginning to draw cause-and-effect arrows between social and economic macrovariables down to the level of the child's brain. That connection is potentially quite powerful."[18] Meaney envisions a science of epigenetics that will show, at the molecular level, the fine-grained biological effects of early stress, deprivation, and trauma and provide support for social policies to reduce these harms. Szyf extends the implications of epigenetics to the unification of the natural and social sciences and the resolution of the nature-nurture problem. As Szyf writes, "Epigenetics will have a dramatic impact on how we understand history, sociology, and political science. If the environment has a role to play in changing your genome, then we've bridged the gap between social processes and biological processes. That will change the way we look at everything."[19]

A 2013 essay in the *Annual Review of Sociology* by Hannah Landecker and Aaron Panofsky embraced this vision, theorizing epigenetics as offering a "renegotiation and reconfiguring of the biological, the social, and their interrelation."[20] Many gender scholars have been similarly drawn to epigenetics. Feminists see potential in epigenetics for forging collaborations with biological scientists to explore the material embodiment of social gender norms, and for rethinking the boundaries of reproductive bodies and responsibilities in ways that acknowledge how bodies are intimately situated in their physical and social environments.[21] Historian and philosopher of biology Evelyn Fox Keller, in her 2010 book, *The Mirage of a Space between Nature and Nurture*, suggests that new discoveries in research fields such as epigenetics represent a postgenomic science that offers a long-sought alternative to determinist, reductionist, gene-centric explanations of heredity. Political scientist Maurizio Meloni likewise regards epigenetics as enacting "a real rupture," "a milestone in the postgenomic landscape," and "a new thought-style," one that abandons the "rigid assumptions" of molecular biology for an understanding of biology as "inextricably entangled with social and environmental factors."[22]

These varied narratives situate recent interest in epigenetics and in fields such as DOHaD as a restorative return of long-neglected and ostracized ideas, a "soft heredity" to counter the twentieth century's "hard heredity."[23] In this telling, the intensity of present interest in maternal effects is simply a high-profile and very legible example of this development.[24] In this book, I have offered a different view.

DOHaD advances a model of human inheritance and development in which the wider social and physical environment can be seen as heritable and as a determinant of future biomedical outcomes via discrete biochemical modifications introduced by the vector of the maternal body. This model is crisply epitomized by an illustration accompanying a *New Yorker* piece on fetal origins research, in which a vector of future developmental outcomes extends outwards from the fetus in the

FIGURE 9.1 Maternal bodies as epigenetic vectors. Illustration from Stephen S. Hall, "Small and Thin," in the *New Yorker*. Credit: Laurent Cilluffo.

woman's abdomen (fig. 9.1).[25] Rather than challenging genetic determinism and biological reductionism, present-day research programs in human epigenetics strategically appropriate and modify these discourses to include a particular conception of the social determinants of health, one that places the maternal-fetal relation at the center.

Otherwise highly discerning analyses of contemporary epigenetics rarely comment on the prominent figure of the maternal body. Anthropologist Jörg Niewöhner, for example, contrasts the "embedded body" envisioned in epigenetic models to the classic Western biomedical model of the body as mechanistic, genetically determined, and autonomous. But close attention to Niewöhner's articulation of this "emergent phenomenon . . . made plausible by environmental epigenetics" reveals certain omissions common to many accounts of the theoretical innovations of epigenetics.[26] As Niewöhner writes, "Epigenetics produces an 'embedded body,' that is, a body that is heavily impregnated by its own past and by the social and maternal environment within which it dwells. It is a body that is imprinted by evolutionary and transgenerational time, by 'early-life' and a body that is highly susceptible to changes in its social and maternal environment. This notion of the body differs significantly from the individual body with its notion of skin-bound self and autonomy. . . . It suggests an altogether different degree of entanglement between body and 'context.'"[27] The embedded body, according to Niewöhner, is ever-changing, entangled, emergent, and imprinted by its environment. Indeed, it appears that the "embedded body" is an archetypal fetus. In turn, the environment is represented by the fuzzy, receding figure of the maternal.

The gendered social and symbolic matrix informing this postgenomic conception of health optimization is powerfully integrated in a cover image chosen by the German newsmagazine *Der Spiegel* for an issue on the new science of epigenetics (fig. 9.2). A nude, white, blonde, model-gorgeous woman emerges from shimmering water that curls around her body in the shape of a DNA helix. In the background glow DNA sequence read-outs. Shaking off her double helix, the woman rises above her "gene pool." Clean, pure, and born again, she gestures upwards toward the heavens, high above the muck of the pond and her genetic read-out. Bold letters read, "Victory over the genes—Smarter, healthier, happier: How we can outsmart our genetic material." As it conveys the promise of epigenetic science for health optimization and shaping human destiny, the image also packages a dense cluster of associations: birth, fetus, and womb; supernatural transgressions of nature, from mermaids to Holly-wood's ghostly feminine apparitions; stock icons Lady Liberty and the Winged Victory; the prolific symbolism of erotic nude bathing figures in Western art; the forward-looking guardians of the future family in pronatalist propaganda posters; and the conventional representations of cleanliness, purity, and aspirational whiteness in cosmetics and naturopathic advertising.

Taking seriously how epigenetics renders certain bodies as carriers of risk and as intensive targets of intervention opens an opportunity to tell unexpected and orthogonal histories of the present molecular life sciences, histories that uproot and transform more conventional nar-

FIGURE 9.2 "Victory over the Genes." Cover of 2010 *Der Spiegel* on epigenetics. Copyright Der Spiegel 32/2020.

ratives. An appreciation of the figuring of the gestational reproductive body in epigenetics suggests that epigenetics does not so much "make plausible" the embedded body. Rather, it fixes the molecular gaze *upon* the embedded body, an already-formed and highly charged entity in the science and politics of maternal-fetal relations, and elevates it to the center of biomedical theory, intervention, and surveillance.[28] In this way, epigenetics does not so much "entangle . . . bodies and contexts." Rather, it brings the "environment"—transduced through the maternal body—into processes of biomedicalization, optimization, and manipulation of life already initiated by the twentieth-century molecular life sciences.[29]

Acknowledgments

I began this book eight years ago. It was written in three distinct periods: before the birth of my children, after the birth of my first child, and after the birth of my second child. My first thanks is to my partner, children, sister, and parents, for their support for my research and writing during this time. My second thanks is to the feminist scholars and activists whose formidable analyses of motherhood, pregnancy, birth, and reproductive justice serve as a departure point for this work.

A number of talented young scholars contributed to this project as research assistants. Working with them was one of the most gratifying aspects of the writing and research process. My sincere thanks to Alex Borsa, Tess Consoli, Bradley Craig, Annika Gompers, Alex Margarite, Amelia Ross, Katelyn Smith, Amanda Villani, and Kate Womersley for their indispensable assistance.

Throughout the writing of this book, I was in continual dialogue with Rene Almeling, Rich Olson, Meredith Reiches, and Alexa Richardson about its central questions. I thank them for their powerful friendship and intellectual generosity along this journey. Conversations with Janet Golden, David Haig, Evelyn Fox Keller, Nancy Krieger, Quill Kukla, Chris Kuzawa, Hannah Landecker, Martine Lappé, Ruth Müller, Dorothy Roberts, Yoel Sadovsky, Gemma Sharp, Tobias Uller, Mike Wade, and Meg

Waren were also profoundly important to my thinking at critical points in the writing of the book. I thank them for their scholarship and for their engagement with my work. Over the years, many people offered encouragement, provided feedback on work in progress, and/or collaborated on elements of this project, including Jere Behrman, Cynthia Daniels, Michael Dietrich, Mary Fissell, Matthew Gillman, Kimberly Hamlin, Rebecca Lemov, Stéphanie Lloyd, Margaret Lock, Maurizio Meloni, Naomi Oreskes, Katy Park, Angela Potochnik, Rayna Rapp, Miriam Rich, Suzanne Richardson, Janet Rich-Edwards, Stacey Ritz, Elizabeth F. S. Roberts, Londa Schiebinger, Linda Schlossberg, Daniel Schönpflug, Heather Shattuck-Heidorn, Myrna Perez Sheldon, Banu Subramaniam, Sari van Anders, Allen Wilcox, Rasmus Winther, Shannon Withycombe, and the anonymous readers for University of Chicago Press.

The writing of this book was scaffolded by many forms of support. Karen Darling and her excellent staff at the University of Chicago Press championed this project and assisted with the production process. Lauren Rubenzahl provided tactful and incisive developmental editing as the writing progressed. Deborah Valdovinos and the staff of the Department of the History of Science and of Studies of Women, Gender, and Sexuality at Harvard offered administrative support crucial to the completion of this book. I am indebted to Debra Nudel for helping me to find the strength, perspective, and confidence to persist. Natalie Beaumont-Smith of Harvard's Office of Work/Life and Andrea Bergmann of the Wissenschaftskolleg's family support team, caretakers Neia Abreu, Aline Damas, Lisa Jensen, Timo Kelichhaus, Rachel Simon Marshall, Zoë Reiches, Linda Thelan, Lou Trajano, Julia Wallner, and Karolina Wilpert, teachers at the Kita im St.-Michaels-Heim, Neighborhood Preschool, and Oxford Street Daycare, and my extended family formed an embracing structure of childcare support for our family during the writing of this book. My heartfelt thanks to all.

Funding for this research was provided by the Radcliffe Institute for Advanced Study, a William F. Milton Grant, a John L. Loeb Associate Professorship, the Institute for Advanced Study (Wissenschaftskolleg) in Berlin, the Faculty of Arts and Sciences of Harvard University, and a Robert Wood Johnson Foundation seed grant from the Harvard Pop Center. A great many audiences heard elements of this work while in development. I remember each visit fondly, and the discussions generated by these events helped shape the final product. My thanks to the hosts and the participants.

I thank Toronto-based artist Amy Campbell-Noseworthy for her original illustrations and all those who provided permission to reprint images included in this volume. Portions of the introduction, chapter 8, and the epilogue to this book are drawn, in revised form, from previously published work: S. S. Richardson et al., "Society: Don't Blame the Mothers," *Nature* 512, no. 7513 (2014): 131–32; S. S. Richardson, "Maternal Bodies in the Postgenomic Order: Gender and the Explanatory Landscape of Epigenetics," in *Postgenomics: Perspectives on Biology after the Genome*, ed. Richardson and Hallam Stevens (Durham: Duke University Press, 2015), 210–31; S. S. Richardson, "Plasticity and Programming: Feminism and the Epigenetic Imaginary," *Signs* 43, no. 1 (2017): 29–52; G. C. Sharp et al., "It's the Mother! How Assumptions about the Causal Primacy of Maternal Effects Influence Research on the Developmental Origins of Health and Disease," *Social Science & Medicine* 213 (2018): 20–27; and G. C. Sharp et al., "Time to Cut the Cord: Recognizing and Addressing the Imbalance of DOHaD Research towards the Study of Maternal Pregnancy Exposures," *Journal of Developmental Origins of Health and Disease* 10, no. 5 (2019): 509–12.

Sarah S. Richardson
ESSEX, CONNECTICUT

Notes

CHAPTER ONE

1. Yehuda et al., "Holocaust Exposure Induced Intergenerational Effects on FKBP5 Methylation"; Yehuda and Bierer, "Transgenerational Transmission of Cortisol and PTSD Risk."

2. Rosner, *Survivor Café*, 6–7.

3. Gluckman and Hanson, *The Fetal Matrix*; Wells, *The Metabolic Ghetto*; Lumey and Vaiserman, *Early Life Nutrition, Adult Health and Development*; Newnham and Ross, *Early Life Origins of Human Health and Disease*; Annie Murphy Paul, *Origins*.

4. Barker et al., "Weight in Infancy and Death from Ischaemic Heart Disease."

5. Stephen S. Hall, "Small and Thin."

6. Lampl, "Obituary for Professor David Barker," 188.

7. Michelle Murphy, *The Economization of Life*.

8. Fraser et al., "Cohort Profile"; Boyd et al., "Cohort Profile"; "Biological Resources"; Overy, Reynolds, and Tansey, *History of the Avon Longitudinal Study of Parents and Children (ALSPAC), c.1980–2000*.

9. Barker et al., "Growth in Utero, Blood Pressure in Childhood and Adult Life, and Mortality from Cardiovascular Disease," 567.

10. Wells, "Maternal Capital and the Metabolic Ghetto," 11; *The Metabolic Ghetto*.

11. Kuzawa, "Fetal Origins of Developmental Plasticity," 5.

12. Ibid.

13. Ibid., 10, 13.

14. Kuzawa, "Why Evolution Needs Development, and Medicine Needs Evolution," 226; Kuzawa, Gluckman, and Hanson, "Developmental Perspectives on the Origins of Obesity," 207.

15. Kuzawa and Sweet, "Epigenetics and the Embodiment of Race." See also Silverstein, "How Racism Is Bad for Our Bodies"; Kowal, "The Promise of Indigenous Epigenetics."

16. Jason Wolf and Michael Wade, "What Are Maternal Effects (and What Are They Not)?," 1107–8.

17. Lewontin, *The Triple Helix*; Keller, *The Century of the Gene*; Jablonka and Lamb, *Evolution in Four Dimensions*; Oyama, *The Ontogeny of Information*.

18. "Prevention of Neural Tube Defects."

19. "Use of Folic Acid for Prevention of Spina Bifida and Other Neural Tube Defects—1983–1991."

20. Crider, Bailey, and Berry, "Folic Acid Food Fortification."

21. On the ecosocial contingency of prenatal maternal effects, see, e.g., Lumey, Stein, and Susser, "Prenatal Famine and Adult Health"; Reiches, "A Life History Approach to Prenatal Supplementation."

22. Warkany, "Manifestations of Prenatal Nutritional Deficiency," 95–96.

23. Warkany, "Experimental Studies on Nutrition in Pregnancy," 608.

24. Warkany and Kalter, "Congenital Malformations," 1049.

25. Gallagher, "Mother's Diet during Pregnancy Alters Baby's DNA"; Hurley, "Grandma's Experiences Leave a Mark on Your Genes"; Costandi, "Pregnant 9/11 Survivors Transmitted Trauma to Their Children."

26. Richardson et al., "Society: Don't Blame the Mothers."

27. Alberts et al., "Rescuing US Biomedical Research from Its Systemic Flaws."

28. Richardson and Stevens, *Postgenomics*.

29. "Begin Before Birth," BeginBeforeBirth.org.

30. Raine, *The Anatomy of Violence*, 264, 272. On race and the new biosocial criminology, see Larregue and Rollins, "Biosocial Criminology and the Mismeasure of Race."

31. Lu and Chow, "An Interview with David Barker." Emerging scientific ideas about the maternal-fetal imprint may have practical consequences for the constitution of identity and kinship in the postgenomic world, as well as for intuitions about where to place agency and responsibility for fetal harm. See, e.g., Burrell and Edozien, "Surrogacy in Modern Obstetric Practice"; D'Alton-Harrison, "Mater Semper Incertus Est"; Payne, "Grammars of Kinship"; Christiansen, "Who Is the Mother?"; lawsuit *M.R & Anor, An tArd Chlaraitheoir & Ors*, High Court of Ireland (2013).

32. Roseboom et al., "Hungry in the Womb," 144.

33. Hamilton, "How A Pregnant Woman's Choices Could Shape A Child's Health."

34. Martha Kenney and Ruth Müller, "Of Rats and Women"; Valdez, "The Redistribution of Reproductive Responsibility"; Warin et al., "Mothers as Smoking Guns."

35. Duden, *Disembodying Women*.

36. McBride, "Thalidomide and Congenital Abnormalities."

37. Lowe, "Effect of Mothers' Smoking Habits on Birth Weight of their Children"; MacMahon, Alpert, and Salber, "Infant Weight and Parental Smoking Habits"; Meyer and Comstock, "Maternal Cigarette Smoking and Perinatal Mortality."

38. Jones and Smith, "Recognition of the Fetal Alcohol Syndrome in Early Infancy."

39. Markens, Browner, and Press, "Feeding the Fetus."

40. Foster, "Miscarriage and Video Display Terminals"; Lieberman, "Video Display Terminals, 1989."

41. Golden, *Message in a Bottle*; Armstrong, *Conceiving Risk, Bearing Responsibility*.

42. Chavkin et al., "Reframing the Debate"; Lynn Paltrow, "Criminal Prosecutions against Pregnant Women."

43. Quoted in Wingerter, "Fetal Protection Becomes Assault on Motherhood."

44. Purdy, "Are Pregnant Women Fetal Containers?," 289.

45. Annas, "Pregnant Women as Fetal Containers," 13; Atwood, *The Handmaid's Tale*.

46. Chasnoff et al., "Cocaine Use in Pregnancy."

47. Roberts, *Killing the Black Body*, 157. See also Dána-Ain Davis, *Reproductive Injustice*.

48. Roberts, *Killing the Black Body*, 18, 21, 8.

49. Frank et al., "Growth, Development, and Behavior in Early Childhood Following Prenatal Cocaine Exposure"; see also Koren et al., "Bias Against the Null Hypothesis."

50. Pollitt, "Fetal Rights," 288, 297–98.

51. Waggoner, *The Zero Trimester*.

52. Gluckman and Pinal, "Glucose Tolerance in Adults after Prenatal Exposure to Famine."

53. Gluckman and Hanson, *The Fetal Matrix*, 53–54.

54. See, e.g., Denbow, "Good Mothering before Birth."

55. Kukla, *Mass Hysteria*, 112.

56. Ibid., 127.

57. Ibid.

58. On "intensive mothering," see Hays, *The Cultural Contradictions of Motherhood*. Susan Markens, C. H. Browner, and Nancy Press' interview-based research demonstrates pregnant women's increasingly "significant degree of accommodation to clinical advice" and "acceptance of the growing emphasis on the nearly exclusive role of maternal responsibility for fetal outcome" (Markens, Browner, and Press, "Feeding the Fetus," 369).

59. "International Association of Diabetes and Pregnancy Study Groups Recommendations on the Diagnosis and Classification of Hyperglycemia in Pregnancy."

60. Farrar et al., "Association between Hyperglycaemia and Adverse Perinatal Outcomes in South Asian and White British Women"; Lawlor, "The Society for Social Medicine John Pemberton Lecture 2011. Developmental Overnutrition"; Lawlor et al., "Maternal Adiposity."

61. Duley and Farrar, "Commentary: But Why Should Women Be Weighed Routinely during Pregnancy?"

62. Richardson and Almeling, "The CDC Risks Its Credibility with New Pregnancy Guidelines."

63. Mamluk et al., "Low Alcohol Consumption and Pregnancy and Childhood Outcomes."

64. Goodwin, *Policing the Womb*.

65. "A Woman's Rights: Parts 1–7."

66. Goodwin, *Policing the Womb*; Paltrow and Flavin, "Arrests of and Forced Interventions on Pregnant Women in the United States, 1973–2005." Fetal harm is also widely criminalized in non-US contexts; see, e.g., Viterna and Bautista, "Pregnancy and the 40-Year Prison Sentence."

67. Meloni, *Political Biology*; Landecker and Panofsky, "From Social Structure to Gene Regulation, and Back"; Gissis and Jablonka, *Transformations of Lamarckism*; Stotz, "The Ingredients for a Postgenomic Synthesis of Nature and Nurture"; Graham, *Lysenko's Ghost*; Carey, *The Epigenetics Revolution*; Peterson, *The Life Organic*; Meloni, "A Postgenomic Body."

68. Oster, *Expecting Better*; Kiefer, "Expecting Science."

CHAPTER TWO

1. Beauvoir, *The Second Sex*, 11.

2. Translated in Geddes and Thomson, *The Evolution of Sex*, 161. From Weismann, "Beiträge zur Naturgeschichte der Daphnoiden."

3. Weismann, "Amphimixis or the Essential Meaning of Conjugation and Sexual Reproduction (1891)," 101.

4. Russett, *Sexual Science*, 159.

5. Though Weismann was undoubtedly aware of nineteenth-century debates over women's and men's equality, his views on such matters are a mystery lost to the historical record. Barring compelling evidence, any attempt to attribute his ground-clearing assertion of the identical and sexless contribution of males and females to heredity to his personal politics would be irresponsible speculation. Notably, unlike contemporaries such as William Brooks, who mobilized theories of the role of male and female in reproduction to argue against women's increased political equality, Weismann was no outspoken antifeminist in his scientific writings. In any case, as I here demonstrate, implicit or explicit beliefs about gender equality would, at best, represent only one among many of the scientific ambitions, theoretical presuppositions, background assumptions, and evidential constraints guiding Weismann's development of the theory of amphimixis.

6. For a definitive intellectual biography, see Churchill, *August Weismann*.

7. Weismann et al., *Essays upon Heredity and Kindred Biological Problems*; Weismann, *The Germ-Plasm*.

8. Shipley, Poulton, and Schönland, "Editors' Preface to the First Edition," viii.

9. Hale, "Of Mice and Men."

10. Churchill, "Weismann: The Pre-eminent Neo-Darwinian."

11. Weismann, "The Continuity of the Germ-Plasm as the Foundation of a Theory of Heredity (1885)," 170.

12. Charles Darwin, *Variation of Animals and Plants under Domestication*.

13. Weismann, "On Heredity (1883)," 77.

14. Weismann, "The Significance of Sexual Reproduction in the Theory of Natural Selection (1886)," 302.

15. Ibid., 304–5.

16. Ibid., 303.

17. Weismann, "Continuity of the Germ-Plasm," 178.

18. Weismann, "Significance of Sexual Reproduction," 304.

19. Weismann, "Continuity of the Germ-Plasm," 178.

20. Weismann, "Remarks on Certain Problems of the Day (1890)," 90.

21. Laubichler and Davidson, "Boveri's Long Experiment"; Boveri, "Ein Geschlechtlich Erzeugter Organismus ohne Mütterliche Eigenschaften"; "An Organism Produced Sexually without Characteristics of the Mother."

22. Weismann, "Remarks on Certain Problems of the Day (1890)," 92.

23. Hertwig, "Beiträge zur Kenntnis der Bildung, Befruchtung und Theilung des thierischen Eies."

24. For accounts of the significance of this experiment, see "Prof. Oscar Hertwig"; Laubichler, "Hertwig, Wilhelm August Oscar."

25. Churchill, "Hertwig, Weismann, and the Meaning of Reduction Division circa 1890."

26. Huet, *Monstrous Imagination*.

27. Persian and Asian texts on maternal impressions express similar ideas, symbiotic with Western texts. See Justin E. H. Smith, "Imagination and the Problem of Heredity in Mechanist Embryology," 82.

28. Paré and Pallister, *On Monsters and Marvels*, 38, 54–55.

29. Ibid., 54–55.

30. Smith, "Imagination and the Problem of Heredity in Mechanist Embryology," 93.

31. Quotations from *René Descartes, Primae Circa Generationem Animalium et nonnulla de Saporibus, ed. Utrecht Universiteitsbibliotheek (1701)*. Cited in Smith, "Imagination and the Problem of Heredity in Mechanist Embryology," 91. Smith translates *"formatrix omnium membrorum exterioirum"* more literally, as "formative agent of all of the exterior members."

32. Lamarck, *Zoological Philosophy*.

33. See Whitehead, *On the Transmission, from Parent to Offspring, of some Forms of Disease, and of Morbid Taints and Tendencies*, 11.

34. See Ackerknecht, "Diathesis"; Olby, "Constitutional and Hereditary Disorders."

35. Strahan, *Marriage and Disease*, 15–16.

36. Ibid.

37. Rosenberg, "The Bitter Fruit."

38. Beall, "Aristotle's Master Piece in America"; Fissell, "Hairy Women and Naked Truths."

39. *Aristotle's Complete Master Piece*. Quoted in Beall, "Aristotle's Master Piece in America," 214.

40. Whitehead, *On the Transmission, from Parent to Offspring*, 18–20.

41. Jessup, "Monstrosities and Maternal Impressions."

42. Weismann, "The Supposed Transmission of Mutilations (1888)," 445.

43. Ibid., 444.

44. Rosenberg, "The Bitter Fruit," 197; see also Diane B. Paul, *Controlling Human Heredity, 1865 to the Present*, 40–41.

45. While there could be paternal impressions at the moment of conception (see chapter 4), on balance, early modern and premodern texts held that the maternal imprint is more influential than the paternal one. Braidotti, "Signs of Wonder and Traces of Doubt," 297.

46. Ibid., 299.

47. Ibid.

48. For earlier critiques of maternal impressions theories, see Blondel, "The Strength of Imagination in Pregnant Women Examined"; Kant, "Determination of the Concept of a Human Race (Bestimmung des Begriffs einer Menschenrace)"; Erasmus Darwin, *Zoonomia: Or, the Laws of Organic Life*, 515–26.

49. Weismann, "The Supposed Transmission of Mutilations (1888)," 444.

50. Ibid., 447.

51. Ibid., 446.

52. Ibid., 447.

53. Ibid., 444.

54. Ibid., 446.

55. Ibid., 445.

56. Ibid., 447.

57. Owen, *On Parthenogenesis*.

58. Ibid., 72–73, 3–4, 39, 68.

59. See Churchill, "August Weismann Embraces the Protozoa," 779.

60. Already in the 1850s, analogies between Protozoa and the gametes of sexually reproducing multicellular organisms abounded, with researchers describing amoebic organisms as egg-like and ciliates as spermatozoid and characterizing Protozoa conjugation in sexual terms. Churchill, "Hertwig, Weismann, and the Meaning of Reduction Division circa 1890," 430.

61. Maupas, "Le rajeunissement karyogamique chez les Cilies."

62. Churchill, "August Weismann Embraces the Protozoa."

63. Ibid., 787.

64. Geddes and Thomson, *The Evolution of Sex*, 166.

65. Aristotle, *Generation of Animals*, 15.

66. Ibid., 111.

67. Ibid., 185. Aristotle's contemporary Galen instead argued that a mixture of male and female elements constitute the embryo, though like Aristotle, Galen also held that the female seed was of a different sort—"scarcer, colder, and wetter"—and, as historian of medicine Katharine Park notes, "by no means equal to the father's in importance." Park, *Secrets of Women*, 142.

68. Delaney, *The Seed and the Soil*, 8; see also Delaney, "The Meaning of Paternity and the Virgin Birth Debate."

69. Leeuwenhoeck, "An Abstract of a Letter from Mr. Anthony Leeuwenhoeck of Delft about Generation by an Animalcule of the Male Seed," 349, original emphasis.

In contrast, ovists believed that the form of the organism resided in the female; nonetheless, they also assumed that the male principle conferred life or animation. William Harvey insisted that the embryo is not a product of "the vital principle of the mother," but simply a preformed potentiality, like an acorn, stimulated to development by a factor in the male. Ovist Charles Bonnet similarly argued that the female provides the complete form of a generic organism. The female contribution "is a miniature man, a horse, a bull, etc. but it is not a certain man, a certain horse, a certain bull, etc," he wrote. Individual characteristics of an organism result from the male nutritive element—what he called the "seminal liquor." Gasking, *Investigations into Generation, 1651–1828*, 56, 124–26; Harvey, *The Works of William Harvey*.

70. On his discovery of the egg, for instance, German comparative embryologist Karl Von Baer "contrasted the 'wonderful moving life in the seminal vesicles of the male frog' with the 'quiet majesty' of the egg in the germinal vesicle of the female animal." As Gasking writes, in a preview of Emily Martin's famous argument in "The Egg and the Sperm," Von Baer "hinted that . . . this contrast . . . was but a repetition of that contrast between the male and female nature which was manifest in other aspects of everyday life." Gasking, *Investigations into Generation, 1651–1828*, 157; Martin, "The Egg and the Sperm."

71. Weismann, "Amphimixis," 107.

72. Weismann, "Remarks on Certain Problems of the Day (1890)," 90.

73. Ibid., 87.

74. Weismann, "Amphimixis," 106.

75. Ibid., 111.

76. Brooks, *The Law of Heredity*, 74.

77. Ibid., 54, 99.

78. Ibid., 103.

79. Darwin, *Variation of Animals and Plants under Domestication*.

80. Brooks, *The Law of Heredity*, 81–82.

81. Ibid., 84–85.

82. Ibid., 242–43.

83. Ibid., 257.

84. Ibid.

85. Ibid.

86. Ibid., 259.

87. Ibid., 266.
88. Ibid., 263.
89. Geddes and Thomson, *The Evolution of Sex*, vi.
90. Ibid., 162.
91. Ibid., 118–19.
92. Ibid., 122.
93. Ibid., 123, 119.
94. Ibid., 124.
95. Ibid., 125.
96. Ibid., 270, 267.
97. Ibid., 270.
98. Ibid.
99. Ibid.
100. Ibid., 268.
101. Ibid., 270.
102. Ibid., 267.
103. Weismann, "Amphimixis," 105.
104. Weismann, "Remarks on Certain Problems of the Day (1890)," 87.
105. Ibid.
106. Weismann, "Amphimixis," 113.
107. Ibid., 101.
108. Ibid., 111.
109. Weismann, "Continuity of the Germ-Plasm," 252–53.
110. Eduard Strasburger came the closest to Weismann's view of heredity as a truly sex-neutral mingling of the nuclear material of two gametes. Though their positions are more ambiguous, Weismann's contemporaries Theodor Boveri, Ernst Haeckel, Oscar Hertwig, Gustav Jäger, and Carl Nägeli also affirmed versions of the idea of identical or roughly equal maternal and paternal nuclear contribution to heredity. Of course, the rediscovery of Mendel's theory of the random segregation of traits into the parental gametes by Bateson and others around 1900 would later add additional grounding to the doctrine of sex equality in heredity.
111. Castle, *On Germinal Transplantation in Vertebrates*. See also Heape, "Preliminary Note on the Transplantation and Growth of Mammalian Ova within a Uterine Foster-Mother"; "Further Note on the Transplantation and Growth of Mammalian Ova within a Uterine Foster-Mother."
112. Wilson, *The Cell in Development and Inheritance*, 302.
113. Conklin, *Heredity and Environment in the Development of Men*, 117.
114. Ibid., 122–23.
115. Castle, *Genetics and Eugenics*, 49.
116. Ibid., 49–50.
117. Wilson, *The Cell in Development and Heredity*, 667.
118. Ibid.
119. Delaney, *The Seed and the Soil*, 3.
120. Delaney, "The Meaning of Paternity and the Virgin Birth Debate," 508.
121. Ibid.
122. McLaren and Michie, "An Effect of the Uterine Environment upon Skeletal Morphology in the Mouse."

123. For histories of the debate over nuclear monopolism and the question of whether the cytoplasm transmits hereditary elements, see Sapp, *Beyond the Gene*; Keller, *The Century of the Gene*; and Harwood, *Styles of Scientific Thought*.

CHAPTER THREE

1. "The Correspondence School of Gospel and Scientific Eugenics (Advertisement)," 32, 66.

2. The school sold pamphlets offering instruction in eugenics and sex education with a focus on stringent guidelines for "purity" practices among young people. Titles of pamphlets with registered 1912 copyrights to the Correspondence School of Gospel and Scientific Eugenics include "Adolescence"; "Eugenics and Sociology"; "Parents and Teachers, How to Tell the Story of Life"; "Public Speakers and Organizers"; "Young Husbands and Wives"; "Young People of Marriageable Age." *Library of Congress.*

3. "Well-Born Children."

4. Ibid.

5. Drake, *What a Young Wife Ought to Know*, 138.

6. Melendy, *Perfect Womanhood for Maidens, Wives, Mothers*, 7.

7. Additon, *Twenty Eventful Years of the Oregon Woman's Christian Temperance Union, 1880–1900*; "Founds School of Gospel and Scientific Eugenics, Mrs. Mary E. Teats Visiting with Friends in Berkeley"; "Prevention Is Their Object," *Los Angeles Herald*; "Report of the National Woman's Christian Temperance Union Thirty-Fifth Annual Convention."

8. Rhoads, "Mrs. Teats at San Jose."

9. Teats, *The Way of God in Marriage*, 123.

10. Eames, *Principles of Eugenics*, 74.

11. "Ben B. Lindsey," *Encyclopedia Britannica Online*.

12. Peitzman, "Forgotten Reformers."

13. "Dr. Winfield Scott Hall to Deliver Five Addresses"; "Winfield Scott Hall 1861–1942"; Winfield Scott Hall, *Sex Training in the Home*; Winfield Scott Hall and Jeannette Winter Hall, *Sexual Knowledge*.

14. Riddell, *A Child of Light, or, Heredity and Prenatal Culture*. First published in 1900, at least three editions followed, the last in 1915.

15. On evangelical rhetoric in eugenic doctrines, see Durst, "Evangelical Engagements with Eugenics, 1900–1940"; Zenderland, "Biblical Biology"; Rosen, *Preaching Eugenics*.

16. J. S. Morton, "Riddell, Newton N."

17. "Maumee Valley Chautauqua, July 30-Aug. 6, Prof. Newton N. Riddell of Chicago, Ill," *Defiance Daily Crescent News*.

18. Riddell, *Heredity and Prenatal Culture*, 164.

19. Ibid., 20–21.

20. Ibid., 26, 91.

21. Ibid., 164.

22. Ibid.

23. Ibid., 179.

24. Ibid., 205.

25. Ibid., 175.

26. Ibid., 215.

27. Chautauquas were semi-secular middlebrow venues for adult education and, somewhat like TED talks today, pleasant encounters with new ideas. Originating in Chautauqua,

New York, in 1874, similar "Lyceum" or tent Chautauqua programs soon sprung up across the United States. Often mounted as for-profit ventures, they were also events of enormous civic pride that brought unusual talent and new ideas to far-flung areas of rural America.

28. "At Mass Meeting, April 5," *Centralia Chronicle*.

29. Alexander, "Our Literary Folks." Prominent nineteenth-century black feminists who appealed to maternal impressions theories as a mode of racial uplift include Pauline Hopkins, Anna Julia Cooper, and Frances Harper. See Kyla C. Schuller, "Taxonomies of Feeling."

30. "Gave Profitable Lecture, Heredity and Pre-Natal Culture Dr. Riddell's Subject."

31. "Psychology of Heredity Theme," *Rock Island Argus*.

32. Riddell claimed to have "visited all the principal cities of America," "consulted with hundreds of educators, physicians, prison wardens, chiefs of police, superintendents of reformatories, orphanages and insane asylums," and "examined the psychology and heredity of several thousand persons, including some five thousand convicts" in producing his study of heredity and prenatal culture (Riddell, *Heredity and Prenatal Culture*, 21–22).

33. "Psychology of Heredity Theme"; "Art of Brain Building," *Washington Post*; Riddell, "Experiments of Elmer Gates."

34. From "Missionary to China Slated for Lecture," *Galveston Daily News*.

35. "Happy Dispositions Bring Good Health," *Cedar Rapids Evening Gazette*.

36. "Gave Profitable Lecture, Heredity and Pre-Natal Culture Dr. Riddell's Subject."

37. Riddell, *Heredity and Prenatal Culture*, 36, 215.

38. "Vernacular science" refers to publicly accessible scientific claims that operate beyond the boundaries of professional scientific strictures. The circulation and uptake of scientific ideas in the vernacular realm is not delinked from elite science. Rather, it forms part of the discursive context and conditions of possibility for scientific knowledge production. Hence, vernacular sciences represent a critical dimension of the history of science. On vernacular science and persistent beliefs in the inheritance of acquired traits in progressive era public culture, see Pandora, "Knowledge Held in Common."

39. Rosenberg, "The Bitter Fruit"; Diane B. Paul, *Controlling Human Heredity, 1865 to the Present*.

40. Halpern, *American Pediatrics*; Hanson, *A Cultural History of Pregnancy*.

41. Bayer, *Maternal Impressions*, 44, 13.

42. Gates, "The Art of Rearing Children."

43. Kirby, *Transmission*, 10.

44. Clymer, *How to Create the Perfect Baby by Means of the Art or Science Generally Known as Stirpiculture*, 22.

45. Kirby, *Transmission*, 9.

46. Ibid.

47. Holbrook, *Homo-Culture*, 96.

48. Ibid., 98.

49. See, e.g., Burbank, *The Training of the Human Plant*, 69.

50. Gates, *The Relations and Development of the Mind and Brain*, 9.

51. Melendy, *Vivilore*, 306.

52. Ibid., 305.

53. See Schuller, "Taxonomies of Feeling," 282; Arni, "The Prenatal."

54. Hilts, "Obeying the Laws of Hereditary Descent," 63.

55. Van Wyhe, "The Diffusion of Phrenology through Public Lecturing."

56. Alan Gribben, "Mark Twain, Phrenology and the 'Temperaments,'" 53.

57. Combe, *The Constitution of Man Considered in Relation to External Objects*. Combe's

text was "fourth in circulation to the most widely read books of the day—the *Bible, Pilgrim's Progress,* and *Robinson Crusoe,*" according to Stephen Tomlinson, "Phrenology, Education and the Politics of Human Nature," 1–2.

58. Combe, *The Constitution of Man Considered in Relation to External Objects,* 196.

59. Hilts, "Obeying the Laws of Hereditary Descent," 69–70.

60. Tomlinson, "Phrenology, Education and the Politics of Human Nature," 22.

61. Ibid.; Angus McLaren, "Medium and Message," 88; Stern, *Heads and Headlines,* 37.

62. Hilts, "Obeying the Laws of Hereditary Descent," 74.

63. Joynt, "Phrenology in New York State."

64. Young, "Orson Squire Fowler."

65. Fowler, *Creative and Sexual Science,* iv–v.

66. Ibid., 746.

67. Ibid., 751.

68. Ibid., 752, original emphasis.

69. Newton, *Pre-natal Culture.*

70. Ibid., 3. Between 1890 and 1910 a similar movement for "puericulture" in France, led by obstetrician and pediatrician Adolphe Pinard, argued for prenatal and maternal care in the service of racial improvement. See Schneider, "Puericulture, and the Style of French Eugenics."

71. Brittan, *Man and His Relations.* On mesmerism and phrenology, see Winter, *Mesmerized.*

72. Newton, *Pre-natal Culture,* 6.

73. Ibid., 9.

74. Ibid., 10.

75. Ibid., 10–11.

76. Ibid., 15.

77. Ibid., 41.

78. Ibid., 44.

79. Ibid.

80. Ibid., 43.

81. Ibid., 58.

82. Ibid., 59.

83. Ibid., 23.

84. Ibid.

85. Ibid., 35.

86. Ibid.

87. Wald et al., "The Federal Children's Bureau: A Symposium," 27.

88. Clark, "Child Culture."

89. Barnard, *Kindergarten and Child Culture Papers.*

90. See, e.g., Mosher, *Child Culture in the Home;* Hannah Whitall Smith, *Child Culture.*

91. Sizer, "Child Culture"; Uncle, "Joyous and Sensitive."

92. Perkins Gilman, "How Home Conditions React Upon the Family," 603–5.

93. Riddell, *Child Culture According to the Laws of Physiological Psychology and Mental Suggestion.*

94. Riddell, "Suggestions Relating to Children"; "G. Stanley Hall," *Encyclopædia Britannica Online.*

95. "News and Notes."

96. Burbank, *The Training of the Human Plant*, 72. See also Pandora, "Knowledge Held in Common." Widely referenced by prenatal culture authors, Burbank, in turn, provided an enthusiastic written endorsement of Teats' prenatal culture efforts.

97. Riddell, *Heredity and Prenatal Culture*, 33.

98. Shannon and Truitt, *Nature's Secrets Revealed*, 242, original emphasis.

99. As the concept of "brain-building" hints, prenatal culturists were influenced by ideas about body building through "physical culture," or the deliberate improvement of the body through specific and targeted exercises. References to and endorsements by physical culture experts can be seen throughout the prenatal culture literature. "Mrs. Mary Ed. Teats . . . believes thoroughly in a theology which teaches the Physical Culture life," wrote Bernarr [sic] Macfadden, editor of *Physical Culture* magazine, in his preface to Teats' 1906 book, endorsing her recommendations for developing men and women of stronger character through proper prenatal conditioning. Teats, *The Way of God in Marriage*, preface by Bernarr Macfadden, xix–xx.

100. Riddell, *Heredity and Prenatal Culture*, 73.

101. Gates, "The Art of Rearing Children," 248.

102. Gates, *The Relations and Development of the Mind and Brain*, 13–14.

103. Bayer, *Maternal Impressions*, 2, 20.

104. Ibid., 227.

105. Ibid., 139.

106. Ibid., 226.

107. Ibid., 140.

108. Gates, *The Relations and Development of the Mind and Brain*, 11.

109. Melendy, *The Science of Eugenics and Sex Life*.

110. Ibid., 38.

111. Melendy, *Vivilore*, 41.

112. Riddell, *Heredity and Prenatal Culture*, 228.

113. Bayer, *Maternal Impressions*, 2, 20.

114. Ibid., introduction.

115. Slocum, *For Wife and Mother*, 13.

116. Clymer, *How to Create the Perfect Baby*, 23.

117. Ibid., 70.

118. Melendy, *Vivilore*.

119. Clymer, *How to Create the Perfect Baby*, 23.

120. Ibid., 15. On the history of the analogy of the womb to a stamp, see de Grazia, "Imprints.

121. Kirby, *Transmission*, 6.

122. Ibid.

123. Ibid., 8.

124. Ibid., 20–21, 10, 63–4, 29.

125. On the use of embryological scientific warrants in the nineteenth century to advance women's liberation, see Hayden, *Evolutionary Rhetoric*.

126. On maternalist feminism, see, e.g., Ladd-Taylor, *Mother-Work*. Ladd-Taylor usefully distinguishes "sentimental" and "progressive" maternalists in nineteenth and early twentieth-century women's movements, employing the term "maternalism" "to denote a specific ideology whose adherents hold (1) that there is a uniquely feminine value system based on care and nurturance, (2) that mothers perform a service to the state by raising citizen-workers;

(3) that women are united across class, race, and nation by their common capacity for motherhood and therefore share a responsibility for all the world's children; and (4) that ideally men should earn a family wage to support their 'dependent' wives and children at home" (p. 3).

127. Pendleton, *The Parents Guide for the Transmission of Desired Qualities to Offspring, and Childbirth Made Easy*, 106.

128. Ibid., 34, 66.

129. Cowan, *The Science of a New Life*, 378.

130. Ibid., 380.

131. Ibid., 199.

132. Riddell, *Heredity and Prenatal Culture*, 240.

133. Ibid., 224–25.

134. Ibid., 231.

135. Drake, *What a Young Wife Ought to Know*, 107.

136. Melendy, *Vivilore*, 310–11.

137. Melendy, *Science of Eugenics and Sex Life*, 482.

138. Bayer, *Maternal Impressions*, 2, 105.

139. Clymer, *How to Create the Perfect Baby*, preface, unpaginated.

140. Wrote Stopes, "As regards to the quality of children obtained from unwilling mothers by coercion, it must surely be apparent that not only the physical heredity, but the mental environment of the mother on the child during the ante-natal period, and of the home on the developing infant and school-child, are vital in the production of the right attitude of the race. What sort of people to we expect to breed from reluctant, coerced or mercenary mothers?" Stopes, "Statement of Dr. Marie Stopes," 247.

141. On the particularities of the blend of progressive reform politics, evangelical Christian doctrine, conservative maternalism, and ideologies of domesticity in the temperance and purity movements, see Bordin, *Woman and Temperance*; Epstein, *The Politics of Domesticity*.

142. "Grab A Husband," *Washington Evening Journal*.

143. Popenoe and Johnson, *Applied Eugenics*.

144. "Maternal Impressions—Belief in Their Existence Is Due to Unscientific Method of Thought," 513, 516.

145. Ibid., 516.

146. Jordan, "Prenatal Influences," 38.

147. Riddell, *Heredity and Prenatal Culture*, 85.

148. Ibid., 254.

149. Ibid.

150. Ibid., 254–55.

151. "Happy Dispositions Bring Good Health," *Cedar Rapids Evening Gazette*.

152. Bayer, *Maternal Impressions*, 2, 10.

153. Ibid., 53, 56.

154. Ibid., 12, 60.

155. Ibid., 11.

156. Ibid., 35.

157. Ibid., 43.

158. Ibid., 196.

159. Eames, *Principles of Eugenics*, 74. There is a longer history of women questioning the scientific integrity and misogyny of dismissive views of women's theories of the maternal-fetal relation. Martha Mears in her *The Midwife's Candid Advice to the Fair Sex: or the Pupil of*

Nature (1797) queried scientists' readiness to deny the experiences of women with respect to the maternal-fetal relation: "it is 'the moderns' (i.e. male practitioners) who 'deny any mysterious consent between the mother and the fetus, because it cannot be explained on mechanical principles. Will they for the same reason deny the reciprocal influence of the mind and body? Would it be arguing like a physician or anatomist to deny the absorbing and filtering powers of the placenta, that medium of intercourse between the mother and child, because our dissecting instruments have not been able to trace, nor our glasses to discover to us in that organ either lymphatics or glands for performing such wonderful operations?'" (Mears, p. 57, quoted in Hanson, *A Cultural History of Pregnancy*, 26).

160. Jessup, "Monstrosities and Maternal Impressions."

161. Evans, "Maternal Impressions."

162. Duncan, "Have Maternal Impressions Any Effect on the Fetus in Utero?" Duncan's article prompted several more case studies in subsequent issues of the *Lancet* (see McHattie, 1900; Allen, 1900; Wilson, 1901; Smith, 1905).

163. Shelly, "Superstition in Teratology with Special Reference to the Theory of Impressionism," 308–9.

164. Ibid., 310.

165. Ibid., 311.

166. Ibid.

CHAPTER FOUR

1. Jordan, "Prenatal Influences."

2. Saleeby, *Parenthood and Race Culture*, 245.

3. Michael F. Guyer, *Being Well-Born*, 194.

4. Dixon, "The Truth about So-Called Maternal Impressions," 425–27, original emphasis.

5. Shelly, "Superstition in Teratology with Special Reference to the Theory of Impressionism."

6. Slemons, *The Prospective Mother*, 67–71. Slemons noted that many of his own patients expressed great anxiety over the possibility of maternal impressions. Genetic science, he argued, can provide the "further evidence" a patient needs to know that the theory is untrue. As Slemons wrote, "It is amazing that any person, even though ignorant of medical teaching, should be inclined to attribute abnormal development to something the mother has seen or heard, thought or dreamt, or otherwise experienced while she was pregnant. Yet unfortunately, many do believe this."

7. West, *Prenatal Care*, 19.

8. Read, *The Mothercraft Manual*, 35, 74.

9. Hubbard, *Facts about Motherhood*, 10.

10. Loudon, *Death in Childbirth*.

11. Wertz and Wertz, *Lying-In*.

12. Cravens, *The Triumph of Evolution*.

13. Conklin, *Heredity and Environment in the Development of Man*, 19, 32–34.

14. Ibid., 362–63, 411, 484–85.

15. Ibid., 362–63, 405, 409, 411, 484–85.

16. Spicer, "'A Nation of Imbeciles'"; Ladd-Taylor, "Eugenics, Sterilisation and Modern Marriage in the USA."

17. Popenoe and Johnson, *Applied Eugenics*, 417.

18. Jordan, *The Blood of the Nation.*

19. Galton, *Hereditary Genius*; Davenport, *Heredity in Relation to Eugenics.*

20. Saleeby, *Parenthood and Race Culture*, 33–34.

21. Rodwell, "Dr Caleb Williams Saleeby," 24.

22. Osiro et al., "August Forel (1848–1931)"; David Lee, "Forel, Auguste-Henri."

23. Later in life, Forel publicly renounced eugenics for its racism and violation of individual freedoms. See Kuechenhoff, "The Psychiatrist Auguste Forel and His Attitude to Eugenics."

24. Forel, *The Hygiene of Nerves and Mind in Health and Disease*, 274.

25. Ibid., 210.

26. Ibid., 113.

27. Ibid., 195.

28. Guyer, *Being Well-Born.*

29. Devlin and Wickey, "'Better Living through Heredity.'"

30. Ibid., 202–3.

31. Guyer, *Being Well-Born*, 336.

32. Ibid., 194.

33. Johnson, "The Direct Action of the Environment."

34. Ibid.

35. Maccauley, "The Supposed Inferiority of First and Second Born Members of Families," 168.

36. Huxley, *Brave New World.*

37. Conklin, *Heredity and Environment in the Development of Man*, 308–9.

38. Huxley may have read of experiments such as those of Loeb, or of O. Hertwig on frog sperm immersed in 0.3 per cent chloral hydrate, which produced abnormal offspring.

39. Winther, "August Weismann on Germ-Plasm Variation."

40. Weismann, "The Continuity of the Germ-Plasm as the Foundation of a Theory of Heredity (1885)," 172.

41. Weismann, "On Heredity (1883)," 104.

42. Ibid., 82; Weismann, "Continuity of the Germ-Plasm," 171; "The Significance of Sexual Reproduction in the Theory of Natural Selection (1886)," 321. Today, the simultaneous operation of environmental influences on the somatic and germ cells, leading to intergenerational hereditary effects, is termed "parallel induction."

43. Weismann, "On Heredity (1883)," 101.

44. Weismann, "Continuity of the Germ-Plasm," 173.

45. Weismann, "On Heredity (1883)," 105.

46. Bowler, *The Eclipse of Darwinism*, 88.

47. Conklin, *Heredity and Environment in the Development of Man*, 314, 336.

48. Forel, *The Hygiene of Nerves and Mind in Health and Disease*, 210.

49. Ibid., 124.

50. Ibid., 136; Semon, *The Mneme.*

51. Forel, *The Hygiene of Nerves and Mind in Health and Disease*, 209.

52. Saleeby, *Parenthood and Race Culture*, 4.

53. Ibid., 116.

54. Ibid., 301.

55. Guyer, *Being Well-Born*, 147.

56. Ibid., 11.

57. Guyer and Smith, "Transmission of Eye-Defects Induced in Rabbits by Means of Lens-Sensitized Fowl-Serum."

58. Mott, *Nature and Nurture in Mental Development*, 32–33; for a biographical profile of Mott, see "Frederick Walker Mott."

59. Mott, *Nature and Nurture in Mental Development*, 92–93.

60. Ibid., 32–33.

61. Saleeby, *Parenthood and Race Culture*, 163–64.

62. Janet Golden, *Message in a Bottle*, 22.

63. Mott, *Nature and Nurture in Mental Development*, 91.

64. Saleeby, *Parenthood and Race Culture*, 238.

65. Golden, *Message in a Bottle*, 24–25.

66. Elizabeth M. Armstrong, *Conceiving Risk, Bearing Responsibility*, 54.

67. Ibid., 28.

68. Forel, *The Hygiene of Nerves and Mind in Health and Disease*, 211–12.

69. Saleeby, *Parenthood and Race Culture*, 245.

70. Ibid., 241.

71. Ibid., 288.

72. Ibid., 254.

73. Guyer, *Being Well-Born*, 182–83.

74. Hayden, *Evolutionary Rhetoric*, 123; citing Morrow, *Social Diseases and Marriage, Social Prophylaxis*.

75. Guyer, *Being Well-Born*, 193–94.

76. Ibid., 332.

77. Marshall, *Syphilology and Venereal Disease*, 366–67.

78. Saleeby, *Parenthood and Race Culture*, 292.

79. Marshall, *Syphilology and Venereal Disease*, 14, 296–97.

80. Ibid., 297.

81. Ibid., 302–3, 309–13.

82. See, e.g., Cowan, *The Science of a New Life*, 171; Shannon and Truitt, *Nature's Secrets Revealed*, 221.

83. Cowan, *The Science of a New Life*, 171.

84. Teats, *The Way of God in Marriage*, 78–79.

85. Ibid., 119.

86. Kirby, *Transmission*.

87. Riddell, *Heredity and Prenatal Culture*, 155–56, 175.

88. Shannon and Truitt, *Nature's Secrets Revealed*, 216.

89. See, e.g., Eames, *Principles of Eugenics*, 80; Teats, *The Way of God in Marriage*, 112; R. Swinburne Clymer, *How to create the perfect baby*, 73, 97–98; Kirby, *Transmission*, 49–50; Bayer, *Maternal Impressions*, 92, 198, 250; Melendy, *The Science of Eugenics and Sex Life*, 353.

90. Kirby, *Transmission*, 11–12, 49–50.

91. Eames, *Principles of Eugenics*, 80.

92. Drake, *What a Young Wife Ought to Know*, 118.

93. Saleeby, *Parenthood and Race Culture*, 224.

94. Ibid., 216.

95. Guyer, *Being Well-Born*, 333.

96. Crackanthorpe, *Population and Progress*, 116.

97. Saleeby, *Parenthood and Race Culture*, 195.

98. Macfadden, *Manhood and Marriage*, 75.

99. West, *Prenatal Care*, 4, 20–21.

100. Read, *The Mothercraft Manual*, 75.

101. Hubbard, *Facts about Motherhood*, 8–10.

102. Ballantyne, *Expectant Motherhood*, xiv.

103. Ibid., 199.

104. Saleeby, *Parenthood and Race Culture*, viii.

105. See, e.g., Crackanthorpe, *Population and Progress*, 95–96.

106. Saleeby, *Parenthood and Race Culture*, 208–9, 118.

107. Forel, *The Hygiene of Nerves and Mind in Health and Disease*, 284–85.

108. Saleeby, *Parenthood and Race Culture*, 118.

109. Ibid., 21.

110. Conklin, *Heredity and Environment in the Development of Man*, 101, 139, 291, 201, 42–43.

111. Ibid., 85, 95.

112. Today, paternal effects and preconception care are newly in vogue and constitute an expanding area of scientific inquiry, albeit still dwarfed by the surging field of research on maternal-fetal origins of health. See Almeling, *Guynecology*; Waggoner, *The Zero Trimester*.

CHAPTER FIVE

1. Biffen, "Mendel's Laws of Inheritance and Wheat Breeding," 4.

2. Ibid., 14.

3. Ibid., 38–39.

4. William Bateson, *Mendel's Principles of Heredity*, 245.

5. Ibid., 258.

6. Ibid., 259.

7. Ibid., 262.

8. Ibid., 264.

9. Toyama, "Maternal inheritance and Mendelism."

10. Ibid., 392.

11. Ibid., 377, 381.

12. Ibid., 377.

13. Terao, "Maternal Inheritance in the Soy Bean," 53.

14. Boycott and Diver, "On the Inheritance of Sinistrality in *Limnaea peregra*"; Boycott et al., "The Inheritance of Sinistrality in *Limnaea peregra* (Mollusca, Pulmonata)"; Gurdon, "Sinistral Snails and Gentlemen Scientists."

15. Boycott et al., "The Inheritance of Sinistrality in *Limnaea peregra* (Mollusca, Pulmonata)," 52.

16. Ibid., 55–56.

17. Ibid., 71–72.

18. Thomas Hunt Morgan, *The Physical Basis of Heredity*, 227–28.

19. Ibid., 228–29.

20. Ibid., 227.

21. Uda, "On 'Maternal Inheritance,'" 323, 328–29.

22. Ibid., 332–33.

23. Sturtevant, "Inheritance of Direction of Coiling in *Limnaea*."

24. Tanaka, "Maternal Inheritance in *Bombyx mori*," 479.

25. Boycott et al., "The Inheritance of Sinistrality in *Limnaea peregra* (Mollusca, Pulmonata)," 69.

26. Ibid., 72.

27. Ibid., 69.

28. Goldschmidt, "The Influence of the Cytoplasm upon Gene-Controlled Heredity."

29. Ibid., 11.

30. Ibid., 6.

31. Loeb, *The Organism as a Whole, from a Physicochemical Viewpoint*, iv.

32. Ibid.

33. Wilson, *The Cell in Development and Inheritance*, 301.

34. Ibid., 301, 327.

35. Ibid., 327.

36. Hagemann, "The Foundation of Extranuclear Inheritance."

37. Terao, "Maternal Inheritance in the Soy Bean," 53.

38. Goldschmidt, "The Influence of the Cytoplasm Upon Gene-Controlled Heredity," 7.

39. Ibid., 5.

40. Ibid., 7–8.

41. Ibid., 13–14.

42. Ibid., 14–16.

43. Dobzhansky, "Maternal Effect as a Cause of the Difference between the Reciprocal Crosses in *Drosophila pseudoobscura*."

44. Ibid., 443.

45. Ibid., 446.

46. Walton and Hammond, "The Maternal Effects on Growth and Conformation in Shire Horse-Shetland Pony Crosses," 312.

47. Slater and Edwards, "John Hammond, 1889–1964."

48. Walton and Hammond, "The Maternal Effects on Growth and Conformation in Shire Horse-Shetland Pony Crosses," 312.

49. See, e.g., Castle, "A Further Study of Size Inheritance in Rabbits, with Special Reference to the Existence of Genes for Size Characters."

50. Walton and Hammond, "The Maternal Effects on Growth and Conformation in Shire Horse-Shetland Pony Crosses," 323.

51. Ibid., 331–32.

52. Ibid., 312.

53. "In-Vitro Fertilization Market Size, Share & Trends Analysis Report"; Maggie Fox, "A Million Babies Have Been Born in the U.S. with Fertility Help."

54. Ole Venge, "Studies of the Maternal Influence on the Birth Weight in Rabbits."

55. Rader, *Making Mice*, 8.

56. Artzt, "Mammalian Developmental Genetics in the Twentieth Century."

57. Rader, *Making Mice*, 215.

58. Ibid., 139–40; Jackson and Little, "The Existence of Non-chromosomal Influence in the Incidence of Mammary Tumors in Mice." Later research by Fekete used embryo transplantation to explore the influences of the maternally provided environment, finding a similar effect, though the study was limited as it did not disambiguate the contributions of intrauterine factors from postnatal nursing. Fekete and Little, "Observations on the Mammary Tumor Incidence of Mice Born from Transferred Ova."

59. Fekete, "Differences in the Effect of Uterine Environment upon Development in the DBA and C57 Black Strains of Mice," 413.

60. Green and Russell, "A Difference in Skeletal Type between Reciprocal Hybrids of Two Inbred Strains of Mice (C57 Blk and C3h)."

61. Ibid., 650.

62. McLaren and Michie, "Factors Affecting Vertebral Variation in Mice," 646.

63. McLaren and Michie, "An Effect of the Uterine Environment upon Skeletal Morphology in the Mouse," 1147.

64. McLaren and Michie, "Factors Affecting Vertebral Variation in Mice," 658.

65. Ibid., 657.

66. Bradford, "The Role of Maternal Effects in Animal Breeding, VII: Maternal Effects in Sheep," 1324.

67. Ibid., 1332–33.

68. Ibid., 1330.

69. Dickerson, "Composition of Hog Carcasses as Influenced by Heritable Differences in Rate and Economy of Gain," 497–98.

70. Ibid., 506–7.

71. Ibid., 521.

72. Ibid., 522.

73. Cundiff, "The Role of Maternal Effects in Animal Breeding, VIII: Comparative Aspects of Maternal Effects."

74. Robison, "The Role of Maternal Effects in Animal Breeding, V: Maternal Effects in Swine," 1303; Falconer, "Maternal Effects and Selection Response."

75. Kempthorne, "The Correlations between Relatives in Random Mating Populations," 74.

76. Eisen, "Mating Design for Estimating Direct and Maternal Genetic Variances and Direct-Maternal Genetic Covariances,"13.

77. McLaren and Michie, "An Effect of the Uterine Environment upon Skeletal Morphology in the Mouse."

78. Willham, "The Role of Maternal Effects in Animal Breeding," 1288, 1291.

79. Kline, Stein, and Susser, *Conception to Birth*, 232.

80. Ibid.

CHAPTER SIX

1. Panofsky, *Misbehaving Science*.

2. Jensen, "How Much Can We Boost IQ and Scholastic Achievement?"

3. For a classic development of this view, see Flynn, "Massive IQ Gains in 14 Nations"; and "Race and IQ."

4. Birch and Gussow, *Disadvantaged Children*, 53.

5. Ibid., 35.

6. Present-day instantiations of ideas about maternal intrauterine effects and the intransigent harms of racism can be seen in mobilizations of fetal epigenetic programming science in the context of the history of slavery and ongoing racial discrimination in the United States. See, e.g., Grossi, "New Avenues in Epigenetic Research about Race"; Kuzawa and Sweet, "Epigenetics and the Embodiment of Race."

7. Digital literature search in the biomedical database PubMed for "birthweight," "birth weight," or "birth size" in title or abstract, 1970–present. June 22, 2017.

8. Identical twin studies have been proposed as one method for answering this question. See Behrman and Rosenzweig, "Returns to Birthweight."

9. Lawrence T. Weaver, "In the Balance," 38.

10. Tanner, *A History of the Study of Human Growth*.

11. See, e.g., Hardy, "Rickets and the Rest."

12. "Low Birth Weight Causes Half of US Infant Deaths," *Atlanta Daily World*; Unger, "Weight at Birth and Its Effect on Survival of the Newborn by Geographic Divisions and Urban Rural Areas."

13. Davies, "Low Birthweight Infants"; Squires, "The Tiny Infants."

14. See, for example, Michelson, "Studies in the Physical Development of Negroes." On theories of differences between black and white women's pregnant and reproductive bodies, see Cooper Owens, *Medical Bondage*.

15. Dodge, "Weight of Colored Infants," 344.

16. Anderson, Brown, and Lyon, "Causes of Prematurity," 534.

17. Bivings, "Racial, Geographic, Annual and Seasonal Variations in Birth Weights," 725.

18. Anderson, Brown, and Lyon, "Causes of Prematurity," 523.

19. Boas, "Changes in the Bodily Form of Descendants of Immigrants."

20. "Truth, Reconciliation, and Transformation."

21. Scott, Jenkins, and Crawford, "Growth and Development of Negro Infants, I: Analysis of Birth Weights of 11,818 Newly Born Infants"; Scott et al., "Growth and Development of Negro Infants, III: Growth during the First Year of Life as Observed in Private Pediatric Practice"; Kessler and Scott, "Growth and Development of Negro Infants, II: Relation of Birth Weight, Body Length and Epiphysial Maturation to Economic Status."

22. Crump et al., "Relation of Birth Weight in Negro Infants to Sex, Maternal Age, Parity, Prenatal Care, and Socioeconomic Status," 680.

23. Ibid., 683.

24. Ibid., 694.

25. Kretchmer, "Ecology of the Newborn Infant," 26–27.

26. Keller, *The Century of the Gene*; Kay, *Who Wrote the Book of Life?*

27. Montagu, *The Biosocial Nature of Man*, 71.

28. Ibid., 50.

29. Ibid., 73.

30. For example, see Naeye, Diener, and Dellinger, "Urban Poverty."

31. Winick, "Food and the Fetus," 80; and *Nutrition and Fetal Development*, 102.

32. Clement A. Smith, "Effects of Maternal Undernutrition upon the Newborn Infant in Holland (1944–1945)"; Zena Stein et al., *Famine and Human Development*.

33. Thomson, "Diet in Pregnancy," 521, 517.

34. Ibid.; McGanity et al., "The Vanderbilt Cooperative Study of Maternal and Infant Nutrition."

35. Yerby, "The Disadvantaged and Health Care"; Shapiro et al., "Further Observations on Prematurity and Perinatal Mortality in a General Population and in the Population of a Prepaid Group Practice Medical Care Plan."

36. Kline, Stein, and Susser, *Conception to Birth*, 223.

37. Osofsky, "Poverty, Pregnancy Outcome, and Child Development," 41; see also Costa, "Race and Pregnancy Outcomes in the Twentieth Century."

38. A series of nomenclature statements by the American Academy of Pediatrics in the late 1960s and early 1970s formally distinguished "retarded intrauterine growth" from "prematurity" and set forth guidelines and metrics for the use of these terms. See, e.g.,

Yerushalmy, "Nomenclature, Duration of Gestation, Birth Weight, Intrauterine Growth"; Tanner, "Standards for Birth Weight or Intra-uterine Growth."

39. Little, "Mother's and Father's Birthweight as Predictors of Infant Birthweight"; see also Morton, "The Inheritance of Human Birth Weight"; Udry et al., "Social Class, Social Mobility, and Prematurity."

40. Thomson, "Diet in Pregnancy," 522.

41. Ibid.

42. Ounsted and Ounsted, "Maternal Regulation of Intra-uterine Growth," 996.

43. Ounsted and Ounsted, On Fetal Growth Rate, 57, 65.

44. Ibid., 72–73.

45. Gruenwald, "Fetal Growth as an Indicator of Socioeconomic Change," 868–69.

46. Gruenwald, "Fetal Deprivation and Placental Insufficiency."

47. Towbin, "Mental Retardation Due to Germinal Matrix Infarction," 160.

48. "Diets Threat to Unborn, Expert Says."

49. Zamenhof and Van Marthens, "Study of Factors Influencing Prenatal Brain Development," 167.

50. Bresler, Ellison, and Zamenhof, "Learning Deficits in Rats with Malnourished Grandmothers," 321.

51. Zamenhof and Van Marthens, "Study of Factors Influencing Prenatal Brain Development," 157; Bresler, Ellison, and Zamenhof, "Learning Deficits in Rats with Malnourished Grandmothers," 321.

52. Birch and Gussow, Disadvantaged, xiii, 7.

53. Ibid., 8, 11, 267.

54. Ibid., 7.

55. Wilcox, "On the Importance—and the Unimportance—of Birthweight," 1234.

56. Lumey et al., "Cohort Profile: the Dutch Hunger Winter Families Study."

57. Zena Stein et al., Famine and Human Development, v.

58. Ibid., 3.

59. Ibid., vii.

60. Ibid., 4, 236; Flynn, "Requiem for Nutrition as the Cause of IQ Gains."

61. Ibid., 519.

62. Aylward et al., "Outcome Studies of Low Birth Weight Infants Published in the Last Decade," 518.

63. Kline, Stein, and Susser, Conception to Birth, 172.

64. Ibid., 167.

65. Rush, Stein, and Susser, "A Randomized Controlled Trial of Prenatal Nutritional Supplementation in New York City," 694.

66. Lumley and Donohue, "Aiming to Increase Birth Weight."

67. Habicht et al., "Relation of Maternal Supplementary Feeding during Pregnancy to Birth Weight and Other Socio-biological Factors."

68. Costa, "Race and Pregnancy Outcomes in the Twentieth Century."

69. For a review of supplementation studies, see Reiches, "A Life History Approach to Prenatal Supplementation."

70. Costa, "Race and Pregnancy Outcomes in the Twentieth Century."

71. Chike-Obi et al., "Birth Weight Has Increased over a Generation."

72. Colen et al., "Maternal Upward Socioeconomic Mobility and Black-White Disparities in Infant Birthweight," 2036.

73. Ibid., 2037.

74. Palloni and Morenoff, "Interpreting the Paradoxical in the Hispanic Paradox."

75. Basso and Wilcox, "Intersecting Birth Weight-Specific Mortality Curves."

76. Ibid.

77. Gillman, "Developmental Origins of Health and Disease," 1849.

78. Currie and Hyson, "Is the Impact of Health Shocks Cushioned by Socioeconomic Status?," 245.

79. Ibid., 247.

80. Wilcox, "On the Importance—and the Unimportance—of Birthweight," 1234.

81. Ibid., 1233.

82. Ibid., 1234–35.

83. Ibid., 1239.

84. David, "Commentary: Birthweights and Bell Curves," 1242.

85. Ibid.

86. Ibid.

87. Ibid.

88. See Larregue and Rollins, "Biosocial Criminology and the Mismeasure of Race"; Roberts, "The Ethics of Biosocial Science."

89. Baedke and Nieves Delgado, "Race and Nutrition in the New World," 9.

CHAPTER SEVEN

1. Richardson and Stevens, *Postgenomics*.

2. Heijmans et al., "Persistent Epigenetic Differences Associated with Prenatal Exposure to Famine in Humans," 17046.

3. Ibid., 17048.

4. Web of Science ISI citation statistics, January 2019.

5. Heijmans et al., "Persistent Epigenetic Differences Associated with Prenatal Exposure to Famine in Humans," 17046.

6. Gluckman and Hanson, *The Fetal Matrix*, 88. The term programming had previously been used by Gunter Dörner in the 1960s and 1970s in the field of developmental neuroscience and by Alan Lucas in the context of infant feeding; Barker extended it to the fetal period. Lucas, "Programming by Early Nutrition in Man" and "Role of Nutritional Programming in Determining Adult Morbidity."

7. Barker, *Mothers, Babies, and Health in Later Life*, 13.

8. Gluckman and Hanson, *The Fetal Matrix*.

9. Langley-Evans, Swali, and McMullen, "Lessons from Animal Models," 258.

10. Centre for Fetal Programming, "About CFP."

11. Albert, "Starting from Inside."

12. Ibid.

13. Langley-Evans, Swali, and McMullen, "Lessons from Animal Models," 258.

14. Ibid., 257. On the biological determinism of early life epigenetic programming hypotheses, see Müller and Samaras, "Epigenetics and Aging Research"; Waggoner and Uller, "Epigenetic Determinism in Science and Society."

15. Gluckman and Hanson, *The Fetal Matrix*, 23.

16. Ibid., 63.

17. Kay, *Who Wrote the Book of Life?*

18. Godfrey-Smith, "On the Theoretical Role of 'Genetic Coding.'"

19. Waggoner and Uller, "Epigenetic Determinism in Science and Society"; Richardson, "Maternal Bodies in the Postgenomic Order"; Keller, *The Century of the Gene*, 82. For debates within DOHaD over the term "fetal programming," see Patrick Bateson and Peter Gluckman, *Plasticity, Robustness, Development and Evolution.*

20. Hamilton, "How a Pregnant Woman's Choices Could Shape a Child's Health."

21. Roizen and Oz, *YOU: Having a Baby.*

22. Ibid., 28–33.

23. Hugh D. Morgan et al., "Epigenetic Inheritance at the Agouti Locus in the Mouse."

24. Lee and Zucker, "Vole Infant Development Is Influenced Perinatally by Maternal Photoperiodic History."

25. Szyf et al., "Maternal Programming of Steroid Receptor Expression and Phenotype through DNA Methylation in the Rat."

26. Longino, *Studying Human Behavior.*

27. VanderWeele, *Explanation in Causal Inference.*

28. Ibid.; Heijmans and Mill, "The Seven Plagues of Epigenetic Epidemiology."

29. Pacchierotti and Spano, "Environmental Impact on DNA Methylation in the Germline," 14. On the rapidly evolving and unsettled state of knowledge of epigenetic co-factors, see Lloyd and Raikhel, "'It Was There All Along.'"

30. Flanagan, "Epigenome-Wide Association Studies (EWAS)"; Heijmans and Mill, "The Seven Plagues of Epigenetic Epidemiology."

31. Mill and Heijmans, "From Promises to Practical Strategies in Epigenetic Epidemiology," 587.

32. See, e.g., Suter et al., "Maternal Tobacco Use Modestly Alters Correlated Epigenome-Wide Placental DNA Methylation and Gene Expression."

33. Ian C. G. Weaver et al., "Epigenetic Programming by Maternal Behavior."

34. Waterland and Jirtle, "Transposable Elements." Results for the dose-dependence of methylation levels for graduated phenotypes are similar in Waterland et al., "Maternal Methyl Supplements Increase Offspring DNA Methylation at Axin Fused."

35. Houseman et al., "DNA Methylation Arrays as Surrogate Measures of Cell Mixture Distribution"; Heijmans and Mill, "The Seven Plagues of Epigenetic Epidemiology," 76; Bibikova, "DNA Methylation Microarrays," 3; Flanagan, "Epigenome-Wide Association Studies (EWAS)," 53–54.

36. T. M. Murphy and J. Mill, "Epigenetics in Health and Disease."

37. Greally, "Human Disease Epigenomics 2.0."

38. Birney, Smith, and Greally, "Epigenome-wide Association Studies and the Interpretation of Disease-Omics."

39. Blackwell, "Pregnancy Stress Linked to Babies with Lower IQs."

40. Carmichael, "Who Says Stress is Bad for You?"

41. Blackwell, "Pregnancy Stress Linked to Babies with Lower IQs."

42. L. Cao-Lei et al., "DNA Methylation Signatures Triggered by Prenatal Maternal Stress Exposure to a Natural Disaster: Project Ice Storm"; "DNA Signature Found in Ice Storm Babies," news release.

43. Mosley, "Feeling Stressed? Then Blame Your Mother."

44. Blackwell, "Pregnancy Stress Linked to Babies with Lower IQs."

45. Karim, "University Study Says Stress Experienced while Pregnant Has Lasting Health Impacts for Children."

46. Cao-Lei et al., "Pregnant Women's Cognitive Appraisal of a Natural Disaster Affects DNA Methylation in Their Children 13 Years Later: Project Ice Storm."

47. Cao-Lei et al., "DNA Methylation Mediates the Impact of Exposure to Prenatal Maternal Stress on BMI and Central Adiposity in Children at Age 13½ Years: Project Ice Storm," 749.

48. Cao-Lei et al., "DNA Methylation Signatures Triggered by Prenatal Maternal Stress Exposure to a Natural Disaster: Project Ice Storm," 8.

49. Cao-Lei et al., "Prenatal Stress and Epigenetics."

50. Derfel, "Prenatal Stress from Extreme Weather Is Having Negative Effect on Babies."

51. Ibid.

52. Cao-Lei et al., "DNA Methylation Mediates the Impact of Exposure to Prenatal Maternal Stress on BMI and Central Adiposity in Children at Age 13½ Years: Project Ice Storm," 754.

53. Ibid., 749; Cao-Lei et al., "DNA Methylation Mediates the Effect of Exposure to Prenatal Maternal Stress on Cytokine Production in Children at Age 13½ Years: Project Ice Storm."

54. J. Rijlaarsdam et al., "An Epigenome-Wide Association Meta-analysis of Prenatal Maternal Stress in Neonates."

55. Yehuda et al., "Holocaust Exposure Induced Intergenerational Effects on FKBP5 Methylation."

56. Yehuda and Bierer, "Transgenerational Transmission of Cortisol and PTSD Risk."

57. Klengel et al., "Allele-Specific FKBP5 DNA Demethylation Mediates Gene-Childhood Trauma Interactions."

58. Yehuda et al., "Holocaust Exposure Induced Intergenerational Effects on FKBP5 Methylation," 377, 379.

59. Gradwohl, *Granddaughters of the Holocaust*; cited in Rosner, *Survivor Café*, 8.

60. Kellermann, "Epigenetic Transmission of Holocaust Trauma," 33, 37.

61. Ibid.

62. Rosner, *Survivor café: the legacy of trauma and the labyrinth of memory*, 8–9, 32, 135.

63. Ibid., 62, 149–50.

64. Yehuda et al., "Holocaust Exposure Induced Intergenerational Effects on FKBP5 Methylation," 379; "Key Findings: Study Finds Epigenetic Changes in Children of Holocaust Survivors."

65. Shulevitz, "The Science of Suffering."

66. Kellermann, "Epigenetic Transmission of Holocaust Trauma," 37–38.

67. Yehuda et al., "Holocaust Exposure Induced Intergenerational Effects on FKBP5 Methylation," 379. Klengel is a coauthor on Yehuda's 2016 study.

68. Ibid., 377.

69. Ibid.

70. Ravelli, Stein, and Susser, "Obesity in Young Men after Famine Exposure in Utero and Early Infancy."

71. Despite frequent repetition today of the alarming statistic of a 50 percent increase in obesity following prenatal nutritional deprivation, no studies have replicated this 1970s finding. As a 2011 review of efforts to find obesity-related and other effects in gestational famine cohorts concluded, "Study findings are still diffuse and conflicting, hampered by limited sample size and chance observations, and should still be considered exploratory and hypothesis generating" (Lumey, Stein, and Susser, "Prenatal Famine and Adult Health").

Researchers in the 1970s actually reported *decreased* rates of obesity among those exposed to famine in the last trimester and first few months of life; only exposure during the first half of pregnancy correlated with increased risk of obesity. The finding of a 50 percent increase in obesity risk among those exposed to famine during early gestation was likely an artifact of the study population and the metric for "obesity" used by researchers at the time. The researchers studied 18-year-old Dutch male military conscripts, a group in which the rate of obesity at the time was only 1.7 percent, a very different baseline prevalence than in many regions of the globe today. Furthermore, the variable of "obesity" was defined in a contextually specific way—as weight-for-height greater than 120 percent of the insurance company standard in the Netherlands at that time. Defined by body mass index of 30 or greater, globally, obesity has tripled among adults between 1975 and 2016; more than one-third of adults in the United States are categorized as obese ("Obesity and Overweight," in *Fact Sheets*, World Health Organization). Simply put, the construct of "obesity" today is an entirely different one than the construct employed in the 1970s Dutch Hunger Winter studies.

72. Aryeh D. Stein et al., "Exposure to Famine during Gestation, Size at Birth, and Blood Pressure at Age 59 y: Evidence from the Dutch Famine"; Rundle et al., "Anthropometric Measures in Middle Age after Exposure to Famine during Gestation: Evidence from the Dutch Famine."

73. For a typical example, see the citation of Lumey and Heijmans' 2008 study to claim that "diet has been shown to influence" IGF2 methylation in Rijlaarsdam et al., "Prenatal Unhealthy Diet, Insulin-Like Growth Factor 2 Gene (IGF2) Methylation, and Attention Deficit Hyperactivity Disorder Symptoms in Youth with Early-Onset Conduct Problems."

74. Heijmans et al., "Persistent Epigenetic Differences Associated with Prenatal Exposure to Famine in Humans," 17046, 17048.

75. Tobi et al., "Prenatal Famine and Genetic Variation Are Independently and Additively Associated with DNA Methylation at Regulatory Loci within IGF2/H19."

76. Tobi et al., "Early Gestation as the Critical Time-Window for Changes in the Prenatal Environment to Affect the Adult Human Blood Methylome."

77. Tobi et al., "DNA Methylation as a Mediator of the Association between Prenatal Adversity and Risk Factors for Metabolic Disease in Adulthood," 5.

78. Ibid.

79. Ibid.

80. Richmond, Relton, and Davey Smith, "What Evidence Is Required to Suggest that DNA Methylation Mediates the Association between Prenatal Famine Exposure and Adulthood Disease?"

81. Greally, "Thread: Why Are We Publishing Uninterpretable #epigenetic association studies in @sciencemagazine?," *Twitter.*

CHAPTER EIGHT

1. Eclarinal et al., "Maternal Exercise during Pregnancy Promotes Physical Activity in Adult Offspring"; Ray, "Inheriting Mom's Exercise Regime"; Reynolds, "Babies Born to Run."

2. Paneth and Susser, "Early Origin of Coronary Heart Disease (the 'Barker Hypothesis')."

3. See also Kramer and Joseph, "Enigma of Fetal/Infant-Origins Hypothesis."

4. Gluckman and Pinal, "Glucose Tolerance in Adults after Prenatal Exposure to Famine."

5. Gage, Munafo, and Davey Smith, "Causal Inference in Developmental Origins of Health and Disease (DOHaD) Research."

6. On causal specificity, see Woodward, "Causation in Biology: Stability, Specificity, and the Choice of Levels of Explanation."

7. Machamer, Darden, and Craver, "Thinking about Mechanisms," 17.

8. Ibid., 21.

9. Woodward, *Making Things Happen.*

10. Waterland and Jirtle, "Transposable Elements."

11. Boyle, Li, and Pritchard, "An Expanded View of Complex Traits," 1182.

12. Heijmans and Mill, "The Seven Plagues of Epigenetic Epidemiology," 76.

13. Mill and Heijmans, "From Promises to Practical Strategies in Epigenetic Epidemiology," 585.

14. Heijmans and Mill, "The Seven Plagues of Epigenetic Epidemiology," 74.

15. Mill and Heijmans, "From Promises to Practical Strategies in Epigenetic Epidemiology," 587.

16. Heijmans and Mill, "The Seven Plagues of Epigenetic Epidemiology," 76.

17. Woodward, *Making Things Happen.*

18. Ibid., 146.

19. Ibid.

20. Machamer, Darden, and Craver, "Thinking about Mechanisms," 22.

21. McGill, "Context Effects in Judgments of Causation," 191.

22. Woodward, "Causation in Biology: Stability, Specificity, and the Choice of Levels of Explanation."

23. Lombrozo, "Causal-Explanatory Pluralism."

24. Lombrozo, "Using Science To Blame Mothers."

25. Winett, Wulf, and Wallack, "Framing Strategies to Avoid Mother-Blame in Communicating the Origins of Chronic Disease."

26. Ibid.

27. Baccarelli, "Conceptual Model of Epigenetic Influence on Obesity Risk"; Institute of Medicine and National Research Council, *Examining a Developmental Approach to Childhood Obesity.*

28. Friedman, "Epigenetic Mechanisms for Obesity Risk."

29. Kooijman et al., "The Generation R Study"; Oken et al., "Cohort Profile: Project Viva."

30. John Wright et al., "Cohort Profile: The Born in Bradford Multi-Ethnic Family Cohort Study."

31. For a table summarizing maternal compared to paternal data collection in these and other cohorts, see Sharp, Lawlor, and Richardson, "It's the Mother! How Assumptions about the Causal Primacy of Maternal Effects Influence Research on the Developmental Origins of Health and Disease."

32. Sharp et al., "Time to Cut the Cord: Recognizing and Addressing the Imbalance of DOHaD Research towards the Study of Maternal Pregnancy Exposures."

33. Sharp, Lawlor, and Richardson, "It's the Mother!"

34. On this point, see also Kenney and Müller, "Of Rats and Women"; Lappé, "The Paradox of Care in Behavioral Epigenetics."

35. Sharp et al., "Time to Cut the Cord."

36. Lawlor et al., "Early Life Predictors of Childhood Intelligence."

37. Korenman, Miller, and Sjaastad, "Long-Term Poverty and Child Development in the United States."

38. Langley et al., "Maternal and Paternal Smoking during Pregnancy and Risk of ADHD Symptoms in Offspring"; Langley et al., "Maternal Smoking during Pregnancy as an Environmental Risk Factor for Attention Deficit Hyperactivity Disorder Behaviour."

39. Lawlor, "The Society for Social Medicine John Pemberton Lecture 2011. Developmental Overnutrition";Yu et al., "Pre-pregnancy Body Mass Index in Relation to Infant Birth Weight and Offspring Overweight/Obesity."

40. Davey Smith and Hemani, "Mendelian Randomization"; Lawlor et al., "Exploring the Developmental Overnutrition Hypothesis Using Parental-Offspring Associations and FTO as an Instrumental Variable"; Richmond et al., "Using Genetic Variation to Explore the Causal Effect of Maternal Pregnancy Adiposity on Future Offspring Adiposity."

41. Barker, "The Fetal and Infant Origins of Adult Disease."

42. Roberts, *Fatal Invention*.

43. Geronimus, "Deep Integration."

44. Joan B. Wolf, *Is Breast Best?*

45. Geronimus, "Deep Integration."

46. Ibid.

47. Ibid.

48. Roberts, "The Ethics of Biosocial Science," 124. On the problem of the disjoint between social categories and what biological measures actually capture, and on concerns about what is at stake for our picture of the social dimensions of health in asserting that methylation levels represent a valid and generalizable measure of social processes and structures as experienced by diverse individuals and populations in the womb, see Darling et al., "Enacting the Molecular Imperative"; Shostak and Moinester, "The Missing Piece of the Puzzle? Measuring the Environment in the Postgenomic Moment."

49. Roberts, "The Ethics of Biosocial Science," 124.

50. Ibid., 126.

51. Ibid. See also Baedke and Nieves Delgado, "Race and Nutrition in the New World."

52. Geronimus, "Deep Integration."

53. Winett, Wulf, and Wallack, "Framing Strategies to Avoid Mother-Blame in Communicating the Origins of Chronic Disease."

54. Carey, *The Epigenetics Revolution*; Francis, *Epigenetics*; Dawson Church, *The Genie in your Genes*; Holt and Paterson, "The Ghost in Your Genes"; Cloud, "Why Your DNA Isn't Your Destiny."

55. Nina Martin and Nina Montagne, "The Last Person You'd Expect to Die in Childbirth"; Dána-Ain Davis, *Reproductive Injustice*; Bergström, "Global Maternal Health and Newborn Health."

56. Macdorman, "Race and Ethnic Disparities in Fetal Mortality, Preterm Birth, and Infant Mortality in the United States."

57. Tran et al., "A Systematic Review of the Burden of Neonatal Mortality and Morbidity in the ASEAN Region"; Goldenberg and McClure, "Maternal, Fetal and Neonatal Mortality."

58. Lassi et al., "Interventions to Improve Neonatal Health and Later Survival"; Oster, *Expecting Better*; N. S. Fox, "Dos and Don'ts in Pregnancy."

59. Milner, Duke, and Bucens, "Reducing Newborn Mortality in the Asia–Pacific Region"; Alkema et al., "Global, Regional, and National Levels and Trends in Maternal Mortality between 1990 and 2015, with Scenario-Based Projections to 2030"; Dána-Ain Davis, *Reproductive Injustice*.

60. Cooper Owens and Fett, "Black Maternal and Infant Health: Historical Legacies of Slavery."

61. Villarosa, "Why America's Black Mothers and Babies Are in a Life-or-Death Crisis."

62. Kuzawa, *Workshop on the Maternal Environment.*

63. Kuh and Ben-Shlomo, "Should We Intervene to Improve Fetal and Infant Growth?" See also Rasmussen, "The 'Fetal Origins' Hypothesis."

64. Kuh and Ben-Shlomo, "Should We Intervene to Improve Fetal and Infant Growth?," 15.

65. Ibid., 1.

66. Friedman, "Epigenetic Mechanisms for Obesity Risk"; Saldaña-Tejeda, "Mitochondrial Mothers of a Fat Nation: Race, Gender and Epigenetics in Obesity Research on Mexican Mestizos."

67. Lawlor, "The Society for Social Medicine John Pemberton Lecture 2011. Developmental Overnutrition."

68. Charuchandra, "Maternal Obesity and Diabetes Linked to Autism in Children."

69. Cameron Scott, "Link between Mom's Diet at Conception and Child's Lifelong Health."

70. "You Are What Your Mum Ate: Obesity Research," *ABC News.*

71. Purnell, "The Nutritional Sins of the Mother . . ."

72. de Assis et al., "High-Fat or Ethinyl-Oestradiol Intake during Pregnancy Increases Mammary Cancer Risk in Several Generations of Offspring"; "Diet of Moms-to-Be Can Up Cancer Risk in Two Generations," *Independent*; Lev, "Why You Should Worry About Grandma's Eating Habits."

73. Parker, "Shamed into Health? Fat Pregnant Women's Views on Obesity Management Strategies in Maternity Care," 27.

74. Ibid., 23–24.

75. Parker and Pausé, "'I'm Just a Woman Having a Baby': Negotiating and Resisting the Problematization of Pregnancy Fatness," 3.

76. Ibid., 1.

77. "Obesity and Overweight," in *Fact Sheets*, World Health Organization.

78. Parker and Pausé, "'I'm Just a Woman Having a Baby,'" 2, 8.

79. Parker, "Mothers at Large: Responsibilizing the Pregnant Self for the 'Obesity Epidemic'"; Parker, "Shamed into Health?"; Parker and Pausé, "'I'm Just a Woman Having a Baby.'" See also McPhail et al., "Wombs at Risk, Wombs as Risk: Fat Women's Experiences of Reproductive Care"; Warin et al., "Telescoping the Origins of Obesity to Women's Bodies."

80. Saldaña-Tejeda, "Mitochondrial Mothers of a Fat Nation."

81. Müller, "A Task That Remains before Us: Reconsidering Inheritance as a Biosocial Phenomenon."

82. Richardson et al., "Society: Don't Blame the Mothers"; Winett, Wulf, and Wallack, "Framing Strategies to Avoid Mother-Blame in Communicating the Origins of Chronic Disease."

83. Weismann, "The Supposed Transmission of Mutilations (1888)," 447.

CHAPTER NINE

1. Costandi, "Pregnant 9/11 Survivors Transmitted Trauma to Their Children."

2. Lauderdale, "Birth Outcomes for Arabic-Named Women in California before and After

September 11"; El-Sayed, Hadley, and Galea, "Birth Outcomes among Arab Americans in Michigan before and after the Terrorist Attacks of September 11, 2001."

3. Berkowitz et al., "The World Trade Center Disaster and Intrauterine Growth Restriction"; Engel et al., "Psychological Trauma Associated with the World Trade Center Attacks and Its Effect on Pregnancy Outcome"; Eskenazi et al., "Low Birthweight in New York City and Upstate New York Following the Events of September 11th."

4. Landrigan et al., "Impact of September 11 World Trade Center Disaster on Children and Pregnant Women." A historical analogue can be found in late nineteenth-century French interest in the influence of "mental shocks" related to the Franco-Prussian War and the *Commune de Paris* in 1870–1871 on pregnant women and their infants in Paris. As historian Caroline Arni relates, Georges Barral "made use of the *enfants du siege* as evidence for his argument that the circumstances of the procreative event had a 'hereditary' relevance." Arni, "The Prenatal: Contingencies of Procreation and Transmission in the Nineteenth Century," 290–91.

5. Hendrickx and Van Hoyweghen have coined the term "biopolitical imputation" to describe appeals to epigenetics to settle questions and direct political attention in social policy domains. Hendrickx and Van Hoyweghen, "Solidarity after Nature: From Biopolitics to Cosmopolitics."

6. Grossi, "New Avenues in Epigenetic Research about Race."

7. Oregon Humanities, "Epigenetics and Equity."

8. Roberts, "The Ethics of Biosocial Science," 121.

9. Krieger, "Embodiment: A Conceptual Glossary for Epidemiology."

10. Roberts, "The Ethics of Biosocial Science," 121.

11. Conley, Strully, and Bennett, *The Starting Gate: Birth Weight and Life Chances.*

12. Ibid., 3.

13. Ibid., 8.

14. Lei et al., "Neighborhood Crime and Depressive Symptoms among African American Women; Reardon, "Poverty Linked to Epigenetic Changes and Mental Illness"; Silverstein, "How Racism Is Bad for Our Bodies."

15. Russell Sage Foundation, "Call for Proposals: Integrating Biology and Social Science Knowledge (BioSS)."

16. Economic and Social Research Council, "Biosocial Research."

17. Hobcraft, "ABCDE of Biosocial Science."

18. Weaver, Meaney, and Szyf, "Maternal Care Effects on the Hippocampal Transcriptome and Anxiety-Mediated Behaviors in the Offspring That Are Reversible in Adulthood."

19. Ibid.

20. Landecker and Panofsky, "From Social Structure to Gene Regulation, and Back," 353.

21. Warin and Hammarström, "Material Feminism and Epigenetics"; Noela Davis, "Politics Materialized: Rethinking the Materiality of Feminist Political Action through Epigenetics"; Malabou, "One Life Only: Biological Resistance, Political Resistance." For critical analysis, see Richardson, "Plasticity and Programming." For explorations of maternal-fetal epigenetics as a theoretical resource for reframing reproductive responsibility, see Lewis, *Full Surrogacy Now*; Michelle Murphy, *The Economization of Life*; Michelle Murphy, "Distributed Reproduction, Chemical Violence, and Latency"; Yoshizawa, "Fetal-Maternal Intra-action"; Lamoreaux, "What If the Environment Is a Person? Lineages of Epigenetic Science in a Toxic China."

22. Meloni, *Political Biology*, 194–95, 203.

23. Ibid.; Graham, *Lysenko's Ghost*; Gissis and Jablonka, *Transformations of Lamarckism*.

24. See, e.g., Meloni, *Political Biology*, 209.

25. Stephen S. Hall, "Small and Thin."

26. Ibid., 290.

27. Ibid.

28. See also Richardson, "Maternal Bodies in the Postgenomic Order"; Landecker, "Food as Exposure."

29. See Clarke et al., "Biomedicalizing Genetic Health, Diseases and Identities."

References

ABC News. "You Are What Your Mum Ate: Obesity Research." March 13, 2012. https://www
.abc.net.au/news/2012-03-14/you-are-what-your-mum-ate3a-obesity-research/3887848.

Ackerknecht, E. H. "Diathesis: The Word and the Concept in Medical History." *Bulletin of the History of Medicine* 56, no. 3 (1982): 317.

Additon, Lucia H. Faxon. *Twenty Eventful Years of the Oregon Woman's Christian Temperance Union, 1880–1900.* Portland: Gotshall, 1904.

Albert, Saige. "Starting from Inside: Fetal Programming Center Looks at Impact of Maternal Diets on Offspring." *Wyoming Livestock Roundup.* December 23, 2017. https://www.wylr
.net/animal-health/301-research/7101-starting-from-inside-fetal-programming-center
-looks-at-impact-of-maternal-diets-on-offspring.

Alberts, Bruce, Marc W. Kirschner, Shirley Tilghman, and Harold Varmus. "Rescuing US Biomedical Research from Its Systemic Flaws." *Proceedings of the National Academy of Sciences* 111, no. 16 (2014): 5773–77.

Alexander, Charles. "Our Literary Folks." *Freeman,* November 24, 1900.

Alkema, Leontine, Doris Chou, Daniel Hogan, Sanqian Zhang, Ann-Beth Moller, Alison Gemmill, Doris Ma Fat, et al. "Global, Regional, and National Levels and Trends in Maternal Mortality between 1990 and 2015, with Scenario-Based Projections to 2030: A Systematic Analysis by the UN Maternal Mortality Estimation Inter-Agency Group." *Lancet* 387, no. 10017 (2016): 462–74.

Almeling, Rene. *Guynecology: Men, Medical Knowledge, and Reproduction.* Oakland: University of California Press, 2020.

Anderson, N. A., E. W. Brown, and R. A. Lyon. "Causes of Prematurity, III: Influence of Race and Sex on Duration of Gestation and Weight at Birth." *American Journal of Diseases of Children* 65, no. 4 (1943): 523–34.

Annas, George. "Pregnant Women as Fetal Containers." *Hastings Center Report* 16 (1986): 3–14.

Aristotle. *Generation of Animals*. Translated by A. L. Peck. Loeb Classical Library. Cambridge, MA: Harvard University Press, 1943.

Aristotle's Complete Master Piece. London: Printed and Sold by the Booksellers, 1762 [multiple editions].

Armstrong, Elizabeth M. *Conceiving Risk, Bearing Responsibility: Fetal Alcohol Syndrome and the Diagnosis of Moral Disorder*. Baltimore: Johns Hopkins University Press, 2003.

Arni, Caroline. "The Prenatal: Contingencies of Procreation and Transmission in the Nineteenth Century." In *Heredity Explored: Between Public Domain and Experimental Science 1850–1930*, edited by Staffan Müller-Wille and Christina Brandt, 285–309. Cambridge, MA: MIT Press, 2016.

Artzt, K. "Mammalian Developmental Genetics in the Twentieth Century." *Genetics* 192, no. 4 (2012): 1151–63.

Atlanta Daily World. "Low Birth Weight Causes Half of US Infant Deaths." March 6, 1975.

Atwood, Margaret. *The Handmaid's Tale*. Toronto: McClelland and Stewart, 1985.

Aylward, Glen P., Steven I. Pfeiffer, Anne Wright, and Steven J. Verhulst. "Outcome Studies of Low Birth Weight Infants Published in the Last Decade: A Metaanalysis." *Journal of Pediatrics* 115, no. 4 (1989): 515–20.

Baccarelli, Andrea. "Conceptual Model of Epigenetic Influence on Obesity Risk." In *Examining a Developmental Approach to Childhood Obesity: The Fetal and Early Childhood Years: Workshop Summary*, edited by Institute of Medicine. Washington, DC: The National Academies, 2015.

Baedke, J., and A. Nieves Delgado. "Race and Nutrition in the New World: Colonial Shadows in the Age of Epigenetics." *Studies in History and Philosophy of Biological and Biomedical Sciences* 76 (2019): 101175.

Ballantyne, J. W. *Expectant Motherhood: Its Supervision and Hygiene*. New York: Funk & Wagnalls, 1914.

Barker, D. J. P. "The Fetal and Infant Origins of Adult Disease." *British Medical Journal* 301, no. 6761 (1990): 1111.

Barker, D. J. P. *Mothers, Babies, and Health in Later Life*. 2nd ed. Edinburgh: Churchill Livingstone, 1998.

Barker, D. J. P. "Rise and Fall of Western Diseases." *Nature* 338, no. 6214 (1989): 371–72.

Barker, D. J. P., J. G. Eriksson, T. Forsen, and C. Osmond. "Fetal Origins of Adult Disease: Strength of Effects and Biological Basis." *International Journal of Epidemiology* 31, no. 6 (2002): 1235–39.

Barker, D. J. P., C. Osmond, J. Golding, D. Kuh, and M. E. Wadsworth. "Growth in Utero, Blood Pressure in Childhood and Adult Life, and Mortality from Cardiovascular Disease." *British Medical Journal* 298, no. 6673 (1989): 564–67.

Barker, D. J. P., P. D. Winter, C. Osmond, B. Margetts, and S. J. Simmonds. "Weight in Infancy and Death from Ischaemic Heart Disease." *Lancet* 2, no. 8663 (1989): 577.

Barnard, Henry. *Kindergarten and Child Culture Papers: Papers on Froebel's Kindergarten, with Suggestions on Principles and Methods of Child Culture in Different Countries*. Syracuse, NY: C. W. Bardeen, 1880.

Basso, O., and A. J. Wilcox. "Intersecting Birth Weight-Specific Mortality Curves: Solving the Riddle." *American Journal of Epidemiology* 169, no. 7 (2009): 787–97.

Bateson, Patrick, and Peter Gluckman. *Plasticity, Robustness, Development and Evolution*. Cambridge: Cambridge University Press, 2011.

Bateson, William. *Mendel's Principles of Heredity*. Cambridge: Cambridge University Press, 1909.

Bayer, C. J. *Maternal Impressions: A Study in Child Life before and after Birth and Their Effect upon Individual Life and Character.* Vol. 2. Winona, MN: Jones & Kroeger, 1897.

Beall, Otho T. "Aristotle's Master Piece in America: A Landmark in the Folklore of Medicine." *William and Mary Quarterly* 20, no. 2 (1963): 207–22.

Beauvoir, Simone de. *The Second Sex.* 1st American ed. New York: Knopf, 1953.

"Begin Before Birth." BeginBeforeBirth.org.

Behrman, Jere R., and Mark R. Rosenzweig. "Returns to Birthweight." *Review of Economics and Statistics* 86, no. 2 (2004): 586–601.

"Ben B. Lindsey." *Encyclopedia Britannica Online.* http://www.britannica.com/biography/Ben-B-Lindsey.

Bergström, Staffan. "Global Maternal Health and Newborn Health: Looking Backwards to Learn from History." *Best Practice & Research Clinical Obstetrics & Gynaecology* 36 (2016): 3–13.

Berkowitz, Gertrude S., Mary S. Wolff, Teresa M. Janevic, Ian R. Holzman, Rachel Yehuda, and Philip J. Landrigan. "The World Trade Center Disaster and Intrauterine Growth Restriction." *JAMA: Journal of the American Medical Association* 290, no. 5 (2003): 595–96.

Bibikova, Marina. "DNA Methylation Microarrays." In *Epigenomics in Health and Disease,* edited by Mario F. Fraga and Agustín F. Fernández, 19–46. Boston: Academic Press, 2016.

Biffen, R. H. "Mendel's Laws of Inheritance and Wheat Breeding." *Journal of Agricultural Science* 1, no. 1 (1905): 4–48.

"Biological Resources." Avon Longitudinal Study of Parents and Children. http://www.bristol.ac.uk/alspac/researchers/our-data/biological-resources/.

Birch, Herbert George, and Joan Dye Gussow. *Disadvantaged Children: Health, Nutrition and School Failure.* New York: Harcourt, Brace & World, 1970.

Birney, E., G. D. Smith, and J. M. Greally. "Epigenome-Wide Association Studies and the Interpretation of Disease-Omics." *PLOS Genetics* 12, no. 6 (2016): e1006105.

Bivings, Lee. "Racial, Geographic, Annual and Seasonal Variations in Birth Weights." *American Journal of Obstetrics and Gynecology* 27 (1934): 725–726.

Blackwell, Tom. "Pregnancy Stress Linked to Babies with Lower IQs: 1998 Quebec Ice-Storm Study." *National Post* (Canada), July 10, 2004, A5.

Blondel, James. *The Power of the Mother's Imagination over the Fœtus Examin'd.* London: John Brotherton, 1729.

Boas, Franz. "Changes in the Bodily Form of Descendants of Immigrants." *American Anthropologist* 14, no. 3 (1912): 530–62.

Bordin, Ruth. *Woman and Temperance: The Quest for Power and Liberty, 1873–1900.* Philadelphia: Temple University Press, 1981.

Boveri, Theodor. "Ein Geschlechtlich Erzeugter Organismus ohne Mütterliche Eigenschaften." *Sitzungsb. Gesellsch. Morph. u. Physiol. München* 5 (1889): 73–80.

Boveri, Theodor. "An Organism Produced Sexually without Characteristics of the Mother." *American Naturalist* 27 (1893): 222.

Bowler, Peter J. *The Eclipse of Darwinism: Anti-Darwinian Evolution Theories in the Decades around 1900.* Baltimore: Johns Hopkins University Press, 1983.

Boycott, A. E., and C. Diver. "On the Inheritance of Sinistrality in *Limnaea peregra.*" *Proceedings of the Royal Society of London Series B, Containing Papers of a Biological Character* 95, no. 666 (1923): 207.

Boycott, A. E., C. Diver, S. L. Garstang, and F. M. Turner. "The Inheritance of Sinistrality in *Limnaea peregra* (Mollusca, Pulmonata)." *Philosophical Transactions of the Royal Society B* 219 (1930): 51–131.

Boyd, Andy, Jean Golding, John Macleod, Debbie A. Lawlor, Abigail Fraser, John Henderson, Lynn Molloy, et al. "Cohort Profile: The 'Children of the 90s'—the Index Offspring of the Avon Longitudinal Study of Parents and Children." *International Journal of Epidemiology* 42, no. 1 (2013): 111–27.

Boyle, E. A., Y. I. Li, and J. K. Pritchard. "An Expanded View of Complex Traits: From Polygenic to Omnigenic." *Cell* 169, no. 7 (2017): 1177–86.

Bradford, G. E. "The Role of Maternal Effects in Animal Breeding, VII: Maternal Effects in Sheep." *Journal of Animal Science* 35, no. 6 (1972): 1324–34.

Braidotti, Rosi. "Signs of Wonder and Traces of Doubt: On Teratology and Embodied Differences." In *Feminist Theory and the Body: A Reader*, edited by Janet Price and Margrit Shildrick, 291–301. New York: Routledge, 1999.

Bresler, David E., Gaylord Ellison, and Stephen Zamenhof. "Learning Deficits in Rats with Malnourished Grandmothers." *Developmental Psychobiology* 8, no. 4 (1975): 315–23.

Brittan, Samuel Byron. *Man and His Relations: Illustrating the Influence of the Mind on the Body; the Relations of the Faculties to the Organs, and to the Elements, Objects and Phenomena of the External World.* New York: W. A. Townsend, 1864.

Brooks, William Keith. *The Law of Heredity: A Study of the Cause of Variation, and the Origin of Living Organisms.* Baltimore: J. Murphy, 1883.

Burbank, Luther. *The Training of the Human Plant.* New York: Century, 1906.

Burrell, Celia, and Leroy C. Edozien. "Surrogacy in Modern Obstetric Practice." *Seminars in Fetal and Neonatal Medicine* 19, no. 5 (2014): 272–78.

Cao-Lei, L., Kelsey N. Dancause, Guillaume Elgbeili, David P. Laplante, Moshe Szyf, and Suzanne King. "DNA Methylation Mediates the Effect of Maternal Cognitive Appraisal of a Disaster in Pregnancy on the Child's C-peptide Secretion in Adolescence: Project Ice Storm." *PLOS One* 13, no. 2 (2018): e0192199.

Cao-Lei, L., Kelsey N. Dancause, Guillaume Elgbeili, Renaud Massart, Moshe Szyf, Aihua Liu, David P. Laplante, and Suzanne King. "DNA Methylation Mediates the Impact of Exposure to Prenatal Maternal Stress on BMI and Central Adiposity in Children at Age 13½ Years: Project Ice Storm." *Epigenetics* 10, no. 8 (2015): 749–61.

Cao-Lei, L., S. R. de Rooij, S. King, S. G. Matthews, G. A. S. Metz, T. J. Roseboom, and M. Szyf. "Prenatal Stress and Epigenetics." *Neuroscience & Biobehavioral Reviews* 117 (2017): 198–210.

Cao-Lei, L., G. Elgbeili, R. Massart, D. P. Laplante, M. Szyf, and S. King. "Pregnant Women's Cognitive Appraisal of a Natural Disaster Affects DNA Methylation in Their Children 13 years Later: Project Ice Storm." *Translational Psychiatry* 5 (2015): e515.

Cao-Lei, L., R. Massart, M. J. Suderman, Z. Machnes, G. Elgbeili, D. P. Laplante, M. Szyf, and S. King. "DNA Methylation Signatures Triggered by Prenatal Maternal Stress Exposure to a Natural Disaster: Project Ice Storm." *PLOS ONE* 9, no. 9 (2014): e107653.

Cao-Lei, L., Franz Veru, Guillaume Elgbeili, Moshe Szyf, David P. Laplante, and Suzanne King. "DNA Methylation Mediates the Effect of Exposure to Prenatal Maternal Stress on Cytokine Production in Children at Age 13½ Years: Project Ice Storm." *Clinical Epigenetics* 8, no. 1 (2016): 54.

Carey, Nessa. *The Epigenetics Revolution: How Modern Biology Is Rewriting Our Understanding of Genetics, Disease, and Inheritance.* New York: Columbia University Press, 2012.

Carmichael, Mary. "Who Says Stress Is Bad for You?" *Newsweek*, February 23, 2009.

Castle, W. E. "A Further Study of Size Inheritance in Rabbits, with Special Reference to the Existence of Genes for Size Characters." *Journal of Experimental Zoology* 53, no. 3 (1929): 421–54.

Castle, William E. *Genetics and Eugenics: A Text-Book for Students of Biology and a Reference Book for Animal and Plant Breeders.* 1st ed. Cambridge, MA: Harvard University Press, 1916.

Castle, William E. *On Germinal Transplantation in Vertebrates.* Edited by John C. Phillips. Washington, DC: Carnegie Institution of Washington, 1911.

Cedar Rapids Evening Gazette. "Happy Dispositions Bring Good Health." November 13, 1913, 3.

Centre for Fetal Programming. "About CFP." https://www.cfp-research.com/about-cfp.

Centralia Chronicle. "At Mass Meeting, April 5." April 1, 1908, 2.

Charuchandra, Sukanya. "Maternal Obesity and Diabetes Linked to Autism in Children." *Scientist,* October 1, 2018.

Chasnoff, Ira J., William J. Burns, Sidney H. Schnoll, and Kayreen A. Burns. "Cocaine Use in Pregnancy." *New England Journal of Medicine* 313, no. 11 (1985): 666–69.

Chavkin, Wendy, Denise Paone, Patricia Friedmann, and Ilene Wilets. "Reframing the Debate: Toward Effective Treatment for Inner City Drug-Abusing Mothers." *Bulletin of the New York Academy of Medicine* 70, no. 1 (1993): 50–68.

Chike-Obi, U., R. J. David, R. Coutinho, and S. Y. Wu. "Birth Weight Has Increased over a Generation." *American Journal of Epidemiology* 144, no. 6 (1996): 563–69.

Christiansen, K. "Who Is the Mother? Negotiating Identity in an Irish Surrogacy Case." *Medicine, Health Care and Philosophy* 18, no. 3 (2015): 317–27.

Church, Dawson. *The Genie in Your Genes: Epigenetic Medicine and the New Biology of Intention.* Santa Rosa, CA: Elite Books, 2007.

Churchill, F. B. *August Weismann: Development, Heredity, and Evolution.* Cambridge, MA: Harvard University Press, 2015.

Churchill, F. B. "August Weismann Embraces the Protozoa." *Journal of the History of Biology* 43, no. 4 (2010): 767–800.

Churchill, F. B. "Hertwig, Weismann, and the Meaning of Reduction Division circa 1890." *Isis* 61, no. 4 (1970): 428.

Churchill, F. B. "Weismann: The Pre-eminent Neo-Darwinian." *Endeavour* 27, no. 2 (2003): 46–47.

Clark, Frederick G. "Child Culture: An Experiment Proposed." *New York Evangelist* 47, no. 18 (1876): 6.

Clarke, Adele E., Janet Shim, Sara Shostak, and Alondra Nelson. "Biomedicalizing Genetic Health, Diseases and Identities." In *Handbook of Genetics and Society: Mapping the New Genomic Era,* edited by Paul Atkinson, Peter Glasner, and Margaret Lock. London: Routledge, 2009.

Cloud, John. "Why Your DNA Isn't Your Destiny." *Time,* January 6, 2010.

Clymer, R. Swinburne. *How to Create the Perfect Baby by Means of the Art or Science Generally Known as Stirpiculture; or, Prenatal Culture and Influence in the Development of a More Perfect Race.* Quakertown, PA: Philosophical Pub. Co., 1902.

Coleman, William. *Biology in the Nineteenth Century: Problems of Form, Function, and Transformation.* New York: Wiley, 1971.

Colen, Cynthia G., Arline T. Geronimus, John Bound, and Sherman A. James. "Maternal Upward Socioeconomic Mobility and Black-White Disparities in Infant Birthweight." *American Journal of Public Health* 96, no. 11 (2006): 2032–39.

Combe, George. *The Constitution of Man Considered in Relation to External Objects*. Boston: Carter and Hendee, 1829.

Conklin, Edwin Grant. *Heredity and Environment in the Development of Man*. Princeton: Princeton University Press, 1915.

Conley, Dalton, Kate W. Strully, and Neil G. Bennett. *The Starting Gate: Birth Weight and Life Chances*. Berkeley: University of California Press, 2003.

Cooper Owens, Deirdre. *Medical Bondage: Race, Gender, and the Origins of American Gynecology*. Athens: University of Georgia Press, 2017.

Cooper Owens, Deirdre, and Sharla M. Fett. "Black Maternal and Infant Health: Historical Legacies of Slavery." *American Journal of Public Health* 109, no. 10 (2019): 1342.

"The Correspondence School of Gospel and Scientific Eugenics (Advertisement)." *Physical Culture* 26, no. 4 (1911): 38a.

Costa, Dora L. "Race and Pregnancy Outcomes in the Twentieth Century: A Long-Term Comparison." *Journal of Economic History* 64, no. 4 (2004): 1056–86.

Costandi, Mo. "Pregnant 9/11 Survivors Transmitted Trauma to Their Children." *Guardian*, September 9, 2011.

Cowan, John. *The Science of a New Life*. New York: Fowler & Wells, 1869.

Crackanthorpe, Montague. *Population and Progress*. London: Chapman & Hall, 1907.

Cravens, Hamilton. *The Triumph of Evolution: American Scientists and the Heredity-Environment Controversy, 1900–1941*. Philadelphia: University of Pennsylvania Press, 1978.

Crider, Krista S., Lynn B. Bailey, and Robert J. Berry. "Folic Acid Food Fortification—Its History, Effect, Concerns, and Future Directions." *Nutrients* 3, no. 3 (2011): 370.

Crump, E. P. "Negroes and Medicine." *Pediatrics* 24, no. 1 (1959): 165.

Crump, E. P., C. P. Horton, J. Masuoka, and D. Ryan. "Relation of Birth Weight in Negro Infants to Sex, Maternal Age, Parity, Prenatal Care, and Socioeconomic Status." *Journal of Pediatrics* 51, no. 6 (1957): 678–97.

Cundiff, Larry V. "The Role of Maternal Effects in Animal Breeding, VIII: Comparative Aspects of Maternal Effects." *Journal of Animal Science* 35, no. 6 (1972): 1335–37.

Currie, Janet, and Rosemary Hyson. "Is the Impact of Health Shocks Cushioned by Socioeconomic Status? The Case of Low Birthweight." *American Economic Review* 89, no. 2 (1999): 245–50.

D'Alton-Harrison, Rita. "Mater Semper Incertus Est: Who's Your Mummy?" *Medical Law Review* 22, no. 3 (2014): 357–83.

Darling, K. W., S. L. Ackerman, R. H. Hiatt, S. S. Lee, and J. K. Shim. "Enacting the Molecular Imperative: How Gene-Environment Interaction Research Links Bodies and Environments in the Post-genomic Age." *Social Science & Medicine (1982)* 155 (2016): 51–60.

Darwin, Charles. *Variation of Animals and Plants under Domestication*. New York: New York University Press, [1868] 1988.

Darwin, Erasmus. *Zoonomia: Or, the Laws of Organic Life*. Dublin: P. Byrne and W. Jones, 1794.

Davenport, Charles Benedict. *Heredity in Relation to Eugenics*. New York: Henry Holt, 1911.

Davey Smith, George, and Gibran Hemani. "Mendelian Randomization: Genetic Anchors for Causal Inference in Epidemiological Studies." *Human Molecular Genetics* 23, no. R1 (2014): R89–98.

David, Richard. "Commentary: Birthweights and Bell Curves." *International Journal of Epidemiology* 30, no. 6 (2001): 1241–43.

Davies, P. A. "Low Birthweight Infants: Immediate Feeding Recalled." *Archives of Disease in Childhood* 66, no. 4 (1991): 551–53.

Davis, Dána-Ain. *Reproductive Injustice: Racism, Pregnancy, and Premature Birth*. New York: New York University Press, 2019.

Davis, Noela. "Politics Materialized: Rethinking the Materiality of Feminist Political Action through Epigenetics." *Women: A Cultural Review* 25, no. 1 (2014): 62–77.

de Assis, Sonia, Anni Warri, M. Idalia Cruz, Olusola Laja, Ye Tian, Bai Zhang, Yue Wang, Tim Hui-Ming Huang, and Leena Hilakivi-Clarke. "High-Fat or Ethinyl-Oestradiol Intake during Pregnancy Increases Mammary Cancer Risk in Several Generations of Offspring." *Nature Communications* 3 (2012): 1053.

Defiance Daily Crescent News. "Maumee Valley Chautauqua, July 30–Aug. 6, Prof. Newton N. Riddell of Chicago, Ill." July 19, 1911, 5.

de Grazia, Margareta. "Imprints: Shakespeare, Gutenberg and Descartes." In *Printing and Parenting in Early Modern England*, edited by Douglas A. Brooks, 29–58. Burlington, VT: Ashgate, 2003.

Delaney, Carol. "The Meaning of Paternity and the Virgin Birth Debate." *Man* 21, no. 3 (1986): 494–513.

Delaney, Carol. *The Seed and the Soil: Gender and Cosmology in Turkish Village Society*. Berkeley: University of California Press, 1991.

Denbow, Jennifer. "Good Mothering before Birth: Measuring Attachment and Ultrasound as an Affective Technology." *Engaging Science, Technology, and Society* 5 (2019): 1–20.

Derfel, Aaron. "Prenatal Stress from Extreme Weather Is Having Negative Effect on Babies." *Montreal Gazette*, January 31, 2009.

Descartes, René. *Primae Circa Generationem Animalium et nonnulla de Saporibus*. Utrecht Universiteitsbibliotheek. 1701.

Devlin, D. S., and C. L. Wickey. "'Better Living through Heredity': Michael F. Guyer and the American Eugenics Movement." *Michigan Academician* 16, no. 2 (1984): 199.

Dickerson, G. E. "Composition of Hog Carcasses as Influenced by Heritable Differences in Rate and Economy of Gain." USDA Bureau of Animal Industry. Ames: Agricultural Experiment Station, Iowa State College of Agriculture and Mechanic Arts, 1947.

"Diets Threat to Unborn, Expert Says." *Hartford Courant*, December 8, 1972, 25.

Dixon, Arch. "The Truth about So-Called Maternal Impressions." *Surgery, Gynecology and Obstetrics* 3 (1906): 424–35.

"DNA Signature Found in Ice Storm Babies—Prenatal Maternal Stress Exposure to Natural Disasters Predicts Epigenetic Profile of Offspring." Canada Newswire Telbec, September 29, 2014.

Dobzhansky, T. "Maternal Effect as a Cause of the Difference between the Reciprocal Crosses in *Drosophila pseudoobscura*." *Proceedings of the National Academy of Sciences* 21, no. 7 (1935): 443–46.

Dodge, C. T. J. "Weight of Colored Infants: Growth during the First Eighteen Months." *American Journal of Physical Anthropology* 10, no. 3 (1927): 337–45.

Drake, Amanda J., and Lincoln Liu. "Intergenerational Transmission of Programmed Effects: Public Health Consequences." *Trends in Endocrinology & Metabolism* 21, no. 4 (2010): 206–13.

Drake, Emma F. A. *What a Young Wife Ought to Know*. Philadelphia: Vir, 1901.

"Dr. Winfield Scott Hall to Deliver Five Addresses." *Stanford Daily*, January 17, 1917, 1.

Duden, Barbara. *Disembodying Women: Perspectives on Pregnancy and the Unborn*. Cambridge, MA: Harvard University Press, 1993.

Duley, Lelia, and Diane Farrar. "Commentary: But Why Should Women Be Weighed Routinely during Pregnancy?" *International Journal of Epidemiology* 36, no. 6 (2007): 1283–84.

Duncan, William. "Have Maternal Impressions Any Effect on the Fetus in Utero?" *Lancet* 2, no. 4027 (1900): 1266.

Durst, Dennis L. "Evangelical Engagements with Eugenics, 1900–1940." *Ethics and Medicine* 18, no. 2 (2002): 45–53.

Eames, Blanche. *Principles of Eugenics: A Practical Treatise*. New York: Moffat, Yard, 1914.

Eclarinal, J. D., S. Zhu, M. S. Baker, D. B. Piyarathna, C. Coarfa, M. L. Fiorotto, and R. A. Waterland. "Maternal Exercise during Pregnancy Promotes Physical Activity in Adult Offspring." *FASEB Journal* 30, no. 7 (2016): 2541–48.

Economic and Social Research Council. "Biosocial Research." https://esrc.ukri.org/research/our-research/biosocial-research/.

Eisen, E. F. "Mating Design for Estimating Direct and Maternal Genetic Variances and Direct-Maternal Genetic Covariances." *Canadian Journal of Genetics and Cytology* 9 (1967): 13–22.

El-Sayed, A., C. Hadley, and S. Galea. "Birth Outcomes among Arab Americans in Michigan before and after the Terrorist Attacks of September 11, 2001." *Ethnicity & Disease* 18, no. 3 (2008): 348–56.

Engel, Stephanie Mulherin, Gertrud S. Berkowitz, Mary S. Wolff, and Rachel Yehuda. "Psychological Trauma Associated with the World Trade Center Attacks and Its Effect on Pregnancy Outcome." *Paediatric and Perinatal Epidemiology* 19, no. 5 (2005): 334–41.

Epstein, Barbara Leslie. *The Politics of Domesticity: Women, Evangelism, and Temperance in Nineteenth-Century America*. Middletown, CT: Wesleyan University Press, 1981.

Eskenazi, Brenda, Amy R. Marks, Ralph Catalano, Tim Bruckner, and Paolo G. Toniolo. "Low Birthweight in New York City and Upstate New York following the Events of September 11th." *Human Reproduction* 22, no. 11 (2007): 3013–20.

Evans, F. A. "Maternal Impressions." *JAMA* 40 (1903): 1519.

Falconer, D. S. "Maternal Effects and Selection Response." In *Genetics Today, Proceedings of the XI International Congress on Genetics*, edited by S. J. Geerts, 763–74. Oxford: Pergamon, 1965.

Farrar, D., L. Fairley, G. Santorelli, D. Tuffnell, T. A. Sheldon, J. Wright, L. van Overveld, and D. A. Lawlor. "Association between Hyperglycaemia and Adverse Perinatal Outcomes in South Asian and White British Women: Analysis of Data from the Born in Bradford Cohort." *Lancet Diabetes & Endocrinology* 3, no. 10 (2015): 795–804.

Fekete, E. "Differences in the Effect of Uterine Environment upon Development in the DBA and C57 Black Strains of Mice." *Anatomical Record* 98, no. 3 (1947): 409–15.

Fekete, E. and C. C. Little. "Observations on the Mammary Tumor Incidence of Mice Born from Transferred Ova." *Cancer Research* 2, no. 8 (1942): 525–30.

Fissell, Mary E. "Hairy Women and Naked Truths: Gender and the Politics of Knowledge in 'Aristotle's Masterpiece.'" *William and Mary Quarterly* 60, no. 1 (2003): 43–74.

Flanagan, J. M. "Epigenome-Wide Association Studies (EWAS): Past, Present, and Future." *Methods in Molecular Biology* 1238 (2015): 51–63.

Flynn, James R. "Massive IQ Gains in 14 Nations: What IQ Tests Really Measure." 101, no. 2 (1987): 171.

Flynn, James R. "Race and IQ: Jensen's Case Refuted." In *Arthur Jensen: Consensus and Controversy*, edited by Sohan Mogdil and Celia Mogdil, 221–32. New York: Falmer, 1987.

Flynn, James R. "Requiem for Nutrition as the Cause of IQ Gains: Raven's Gains in Britain 1938–2008." *Economics & Human Biology* 7, no. 1 (2009): 18–27.

Forel, August. *The Hygiene of Nerves and Mind in Health and Disease*. Translated by Herbert Austin Aikins. 2nd ed. New York: Putnam's Sons, 1907.

Foster, Kenneth R. "Miscarriage and Video Display Terminals: An Update." In *Phantom Risk: Scientific Inference and the Law*, edited by Kenneth R. Foster, David E. Bernstein and Peter W. Huber, 123–37. Boulder: NetLibrary, 1999.

"Founds School of Gospel and Scientific Eugenics, Mrs. Mary E. Teats Visiting with Friends in Berkeley." *San Francisco Call*, 1909.

Fowler, O. S. *Creative and Sexual Science*. Philadelphia: National Pub. Co, 1870.

Fox, Maggie. "A Million Babies Have Been Born in the U.S. with Fertility Help." NBC News, April 28, 2017.

Fox, N. S. "Dos and Don'ts in Pregnancy: Truths and Myths." *Obstetrics & Gynecology* 131, no. 4 (2018): 713–21.

Francis, Richard C. *Epigenetics: The Ultimate Mystery of Inheritance*. New York: W. W. Norton, 2011.

Frank, D. A., M. Augustyn, W. Knight, T. Pell, and B. Zuckerman. "Growth, Development, and Behavior in Early Childhood Following Prenatal Cocaine Exposure: A Systematic Review." *JAMA* 285, no. 12 (2001): 1613–25.

Fraser, A., C. Macdonald-Wallis, K. Tilling, A. Boyd, J. Golding, G. Davey Smith, J. Henderson, et al. "Cohort Profile: The Avon Longitudinal Study of Parents and Children: ALSPAC Mothers Cohort." *International Journal of Epidemiology* 42, no. 1 (2013): 97–110.

"Frederick Walker Mott." *Journal of Nervous and Mental Disease* 64, no. 5 (1926): 555–59.

Friedman, Jacob E. "Epigenetic Mechanisms for Obesity Risk." In *Examining a Developmental Approach to Childhood Obesity: The Fetal and Early Childhood Years: Workshop Summary*. Washington, DC: National Academies, 2015.

Gage, S. H., M. R. Munafo, and G. Davey Smith. "Causal Inference in Developmental Origins of Health and Disease (DOHaD) Research." *Annual Review of Psychology* 67 (2016): 567–85.

Gallagher, James. "Mother's Diet during Pregnancy Alters Baby's DNA." BBC News, April 18, 2011. https://www.bbc.com/news/health-13119545.

Galton, Francis. *Hereditary Genius: An Inquiry into Its Laws and Consequences*. London: Macmillan, 1869.

Galveston Daily News. "Missionary to China Slated for Lecture." 1914, 16.

Gasking, Elizabeth B. *Investigations into Generation, 1651–1828*. London: Hutchinson, 1967.

Gates, Elmer. "The Art of Rearing Children." *Practical Medicine* 8, no. 5 (1897): 241.

Gates, Elmer. *The Relations and Development of the Mind and Brain*. New York: Theosophical Society, 1909.

"Gave Profitable Lecture, Heredity and Pre-natal Culture Dr. Riddell's Subject." *Iola Register*, December 7, 1917.

Geddes, Patrick, and John Arthur Thomson. *The Evolution of Sex*. London: Walter Scott, 1889.

Geronimus, Arline T. "Deep Integration: Letting the Epigenome Out of the Bottle without Losing Sight of the Structural Origins of Population Health." *American Journal of Public Health* 103 (2013): S56.

Gillman, Matthew W. "Developmental Origins of Health and Disease." *New England Journal of Medicine* 353, no. 17 (2005): 1848–50.

Gissis, Snait, and Eva Jablonka. *Transformations of Lamarckism: From Subtle Fluids to Molecular Biology*. Cambridge, MA: MIT Press, 2011.

Gluckman, Peter, and Mark Hanson. *The Fetal Matrix: Evolution, Development, and Disease.* New York: Cambridge University Press, 2005.

Gluckman, Peter, and C. Pinal. "Glucose Tolerance in Adults after Prenatal Exposure to Famine." *Lancet* 357, no. 9270 (2001): 1798.

Godfrey-Smith, Peter. "On the Theoretical Role of 'Genetic Coding.'" *Philosophy of Science* 67 (2000).

Golden, Janet. *Message in a Bottle: The Making of Fetal Alcohol Syndrome.* Cambridge, MA: Harvard University Press, 2005.

Goldenberg, Robert L., and Elizabeth M. McClure. "Maternal, Fetal and Neonatal Mortality: Lessons Learned from Historical Changes in High Income Countries and Their Potential Application to Low-Income Countries." *Maternal Health, Neonatology and Perinatology* 1, no. 1 (2015): 3.

Goldschmidt, Richard. "The Influence of the Cytoplasm upon Gene-Controlled Heredity." *American Naturalist* 68, no. 714 (1934): 5–23.

Goodwin, Michele. *Policing the Womb: Invisible Women and the Criminal Costs of Motherhood.* New York: Cambridge University Press, 2020.

Gould, Stephen Jay. *The Mismeasure of Man.* New York: Norton, 1981.

"Grab A Husband." *Washington Evening Journal,* October 24, 1907, 2.

Gradwohl, Nirit. *Granddaughters of the Holocaust: Never Forgetting What They Didn't Experience.* Brighton, MA: Academic Studies Press, 2012.

Graham, Loren R. *Lysenko's Ghost: Epigenetics and Russia.* Cambridge, MA: Harvard University Press, 2016.

Greally, John. "Human Disease Epigenomics 2.0." *PLOS Biologue* (2015). http://blogs.plos.org/biologue/2015/07/07/human-disease-epigenomics-2-0/.

Greally, John. "Thread: Why Are We Publishing Uninterpretable #epigenetic association studies in @sciencemagazine?" *Twitter,* January 31, 2018. https://twitter.com/EpgntxEinstein/status/958887480782077952.

Green, E. L., and W. L. Russell. "A Difference in Skeletal Type between Reciprocal Hybrids of Two Inbred Strains of Mice (C57 Blk and C3h)." *Genetics* 36, no. 6 (1951): 641–51.

Gribben, Alan. "Mark Twain, Phrenology and the 'Temperaments': A Study of Pseudoscientific Influence." *American Quarterly* 24, no. 1 (1972): 45–68.

Grossi, Élodie. "New Avenues in Epigenetic Research about Race: Online Activism around Reparations for Slavery in the United States." *Social Science Information* 59, no. 1 (2020): 93–116.

Gruenwald, P. "Fetal Growth as an Indicator of Socioeconomic Change." *Public Health Reports* 83, no. 10 (1968): 867–72.

Gruenwald, P. "Fetal Deprivation and Placental Insufficiency." *Obstetrics and Gynecology* 37, no. 6 (1971): 906–8.

"G. Stanley Hall." *Encyclopedia Britannica Online.* http://www.britannica.com/biography/G-Stanley-Hall.

Gurdon, J. B. "Sinistral Snails and Gentlemen Scientists." *Cell* 123, no. 5 (2005): 751–53.

Guyer, Michael F. *Being Well-Born: An Introduction to Eugenics.* Indianapolis: The Bobbs-Merrill Company, 1916.

Guyer, Michael F., and E. A. Smith. "Transmission of Eye-Defects Induced in Rabbits by Means of Lens-Sensitized Fowl-Serum." *Proceedings of the National Academy of Sciences* 6, no. 3 (1920): 134–36.

Habicht, J. P., C. Yarbrough, A. Lechtig, and R. E, Klein. "Relation of Maternal Supplementary Feeding during Pregnancy to Birth Weight and Other Socio-biological Factors." In *Nutrition and Fetal Development*, edited by Myron Winick, 127–45. New York: Wiley-Interscience, 1974.

Hagemann, Rudolf. "The Foundation of Extranuclear Inheritance: Plastid and Mitochondrial Genetics." *Molecular Genetics and Genomics* 283, no. 3 (2010): 199–209.

Hale, Piers J. "Of Mice and Men: Evolution and the Socialist Utopia. William Morris, H. G. Wells, and George Bernard Shaw." *Journal of the History of Biology* 43, no. 1 (2010): 17–66.

Hall, Stephen S. "Small and Thin: The Controversy over the Fetal Origins of Adult Health." *New Yorker*, November 19, 2007, 52–57.

Hall, Winfield Scott. *Sex Training in the Home.* Chicago: W. E. Richardson, 1914.

Hall, Winfield Scott, and Jeannette Winter Hall. *Sexual Knowledge.* Philadelphia: International Bible House, 1913.

Halpern, Sydney A. *American Pediatrics: The Social Dynamics of Professionalism, 1880–1980.* Berkeley: University of California Press, 1988.

Hamilton, Jon. "How a Pregnant Woman's Choices Could Shape a Child's Health." National Public Radio, September 23, 2013. https://www.npr.org/sections/health-shots/2013/09/23/224387744/how-a-pregnant-womans-choices-could-shape-a-childs-health?t=1551640431874.

Hammond, John. *Farm Animals: Their Breeding, Growth, and Inheritance.* 3rd ed. London: E. Arnold, 1960.

Hanson, Clare. *A Cultural History of Pregnancy: Pregnancy, Medicine, and Culture, 1750–2000.* New York: Palgrave Macmillan, 2004.

Hardy, A. "Rickets and the Rest: Child-Care, Diet and the Infectious Children's Diseases, 1850–1914." *Social History of Medicine* 5, no. 3 (1992): 389–412.

Harvey, William. *The Works of William Harvey.* Edited by Robert Willis. Philadelphia: University of Pennsylvania Press, 1989.

Harwood, Jonathan. *Styles of Scientific Thought: The German Genetics Community, 1900–1933.* Chicago: University of Chicago Press, 1993.

Hayden, Wendy. *Evolutionary Rhetoric: Sex, Science, and Free Love in Nineteenth-Century Feminism.* Carbondale: Southern Illinois University Press, 2013.

Hays, Sharon. *The Cultural Contradictions of Motherhood.* New Haven: Yale University Press, 1996.

Heape, Walter. "Further Note on the Transplantation and Growth of Mammalian Ova within a Uterine Foster-Mother." *Proceedings of the Royal Society of London* 62 (1897): 178–83.

Heape, Walter. "Preliminary Note on the Transplantation and Growth of Mammalian Ova within a Uterine Foster-Mother." *Proceedings of the Royal Society of London* 48 (1890): 457–58.

Heijmans, Bastiaan T., and J. Mill. "The Seven Plagues of Epigenetic Epidemiology." *International Journal of Epidemiology* 41, no. 1 (2012): 74–78.

Heijmans, Bastiaan T., Elmar W. Tobi, Aryeh D. Stein, Hein Putter, Gerard J. Blauw, Ezra S. Susser, P. Eline Slagboom, and L. H. Lumey. "Persistent Epigenetic Differences Associated with Prenatal Exposure to Famine in Humans." *Proceedings of the National Academy of Sciences* 105, no. 44 (2008): 17046–49.

Hendrickx, K., and I. Van Hoyweghen. "Solidarity after Nature: From Biopolitics to Cosmopolitics." *Health* 24, no. 2 (2020): 203–19.

Hertwig, Oscar. "Beiträge zur Kenntnis der Bildung, Befruchtung und Theilung des thierischen Eies." *Morphologisches Jahrbuch* 1 (1876): 347–434.

Hertwig, Oscar. "The Growth of Biology in the Nineteenth Century." In *Address before Congress of Scientists*, 461–78. Aachen: Smithsonian, 1901.

Hilts, Victor L. "Obeying the Laws of Hereditary Descent: Phrenological Views on Inheritance and Eugenics." *Journal of the History of the Behavioral Sciences* 18, no. 1 (1982): 62–77.

Hobcraft, John. "ABCDE of Biosocial Science." *Society Now: ESRC Research Making an Impact*, no. 24 (Spring 2016): 18–19.

Holbrook, M. L. *Homo-culture; or, the Improvement of Offspring through Wiser Generation*. New York: M. L. Holbrook, 1899.

Holt, Sarah, and Nigel Paterson. "The Ghost in Your Genes." *NOVA*. Boston: WGBH, 2008.

Horton, C. P., and E. P. Crump. "Skin Color in Negro Infants and Parents—Its Relationship to Birth Weight, Reflex Maturity, Socioeconomic Status, Length of Gestation, and Parity." *Journal of Pediatrics* 52, no. 5 (1958): 547–58.

Horton, Richard. "Offline: COVID-19 Is Not a Pandemic." *Lancet* 396, no. 10255 (2020): 874.

Houseman, Eugene Andres, William P. Accomando, Devin C. Koestler, Brock C. Christensen, Carmen J. Marsit, Heather H. Nelson, John K. Wiencke, and Karl T. Kelsey. "DNA Methylation Arrays as Surrogate Measures of Cell Mixture Distribution." *BMC Bioinformatics* 13, no. 1 (2012): 86.

Hubbard, S. Dana. *Facts about Motherhood*. New York: Claremont, 1922.

Huet, Marie. *Monstrous Imagination*. Cambridge, MA: Harvard University Press, 1993.

Hurley, Dan. "Grandma's Experiences Leave a Mark on Your Genes." *Discover*, 2013.

Huxley, Aldous. *Brave New World: A Novel*. London: Chatto & Windus, 1932.

Independent. "Diet of Moms-to-Be Can Up Cancer Risk in Two Generations." April 22, 2010.

Institute of Medicine and National Research Council. *Examining a Developmental Approach to Childhood Obesity: The Fetal and Early Childhood Years: Workshop in Brief*. Edited by Leslie A. Pray. Washington, DC: National Academies Press, 2015. doi:10.17226/21716.

"International Association of Diabetes and Pregnancy Study Groups Recommendations on the Diagnosis and Classification of Hyperglycemia in Pregnancy." *Diabetes Care* 33, no. 3 (2010): 676–82.

"In-Vitro Fertilization Market Size, Share & Trends Analysis Report by Instrument (Disposable Devices, Culture Media, Capital Equipment), by Procedure Type, by End Use, by Region, and Segment Forecasts, 2020–2027." Grand View Research, 2020. https://www.grandviewresearch.com/industry-analysis/in-vitro-fertilization-market.

Jablonka, Eva, and Marion J. Lamb. *Evolution in Four Dimensions: Genetic, Epigenetic, Behavioral, and Symbolic Variation in the History of Life*. Cambridge, MA: MIT Press, 2005.

Jackson, R. B., and C. C. Little. "The Existence of Non-chromosomal Influence in the Incidence of Mammary Tumors in Mice." *Science* 78, no. 2029 (1933): 465–66.

Jensen, Arthur R. "How Much Can We Boost IQ and Scholastic Achievement?" *Harvard Education Review* 39 (1969): 1–123.

Jessup, Jr., R. B. "Monstrosities and Maternal Impressions." *JAMA* 11 (1888): 519–20.

Johnson, Roswell. "The Direct Action of the Environment." *American Breeders Association Proceedings*, no. 5 (1909): 228.

Jones, Kenneth L., and David W. Smith. "Recognition of the Fetal Alcohol Syndrome in Early Infancy." *Lancet* 2 (1973): 999–1001.

Jordan, David Starr. *The Blood of the Nation: A Study of the Decay of Races through Survival of the Unfit*. Boston: American Unitarian Association, 1902.

Jordan, David Starr. "Prenatal Influences." *Journal of Heredity* 5, no. 1 (1914): 38–39.

Joynt, R. J. "Phrenology in New York State." *New York State Journal of Medicine* 73, no. 19 (1973): 2382.

Kant, Immanuel. "Determination of the Concept of a Human Race (Bestimmung des Begriffs einer Menschenrace)." In *Kant and the Concept of Race: Late Eighteenth-Century Writings*, edited by Jon M. Mikkelsen, 125–39. Albany: State University of New York Press, [1785] 2013.

Karim, Malika. "University Study Says Stress Experienced while Pregnant Has Lasting Health Impacts for Children." *Global News*, September 13, 2018.

Kay, Lily E. *Who Wrote the Book of Life? A History of the Genetic Code*. Stanford, CA: Stanford University Press, 2000.

Keller, Evelyn Fox. *The Century of the Gene*. Cambridge, MA: Harvard University Press, 2000.

Keller, Evelyn Fox. *The Mirage of a Space between Nature and Nurture*. Durham, NC: Duke University Press, 2010.

Kellermann, N. P. "Epigenetic Transmission of Holocaust Trauma: Can Nightmares Be Inherited?" *Israel Journal of Psychiatry and Related Sciences* 50, no. 1 (2013): 33–39.

Kempthorne, Oscar. "The Correlations between Relatives in Random Mating Populations." *Cold Spring Harbor Symposia on Quantitative Biology* 20 (1955): 60–78.

Kenney, Martha, and Ruth Müller. "Of Rats and Women: Narratives of Motherhood in Environmental Epigenetics." *BioSocieties* 12, no. 1 (2017): 23–46.

Kessler, A., and R. B. Scott. "Growth and Development of Negro Infants, II: Relation of Birth Weight, Body Length and Epiphysial Maturation to Economic Status." *AMA American Journal of Diseases of Children* 80, no. 3 (1950): 370–78.

"Key Findings: Study Finds Epigenetic Changes in Children of Holocaust Survivors." *Research Currents: Research News from the US Department of Veterans Affairs*. October 20, 2016. https://www.research.va.gov/currents/1016–3.cfm.

Kiefer, Amy. "Expecting Science: Evidence-Based Info for the Thinking Parent." https://expectingscience.com/.

Kirby, Georgiana Bruce. *Transmission; or, Variation of Character through the Mother*. New York: Fowler and Wells, 1882.

Klengel, T., D. Mehta, C. Anacker, M. Rex-Haffner, J. C. Pruessner, C. M. Pariante, T. W. Pace, et al. "Allele-Specific FKBP5 DNA Demethylation Mediates Gene-Childhood Trauma Interactions." *Nature Neuroscience* 16, no. 1 (2013): 33–41.

Kline, Jennie, Zena Stein, and Mervyn Susser. *Conception to Birth: Epidemiology of Prenatal Development*. New York: Oxford, 1989.

Kooijman, Marjolein N., Claudia J. Kruithof, Cornelia M. van Duijn, Liesbeth Duijts, Oscar H. Franco, Marinus H. van IJzendoorn, Johan C. de Jongste, et al. "The Generation R Study: Design and Cohort Update 2017." *European Journal of Epidemiology* 31, no. 12 (2016): 1243–64.

Koren, Gideon, Heather Shear, Karen Graham, and Tom Einarson. "Bias against the Null Hypothesis: The Reproductive Hazards of Cocaine." *Lancet* 334, no. 8677 (1989): 1440–42.

Korenman, Sanders, Jane E. Miller, and John E. Sjaastad. "Long-Term Poverty and Child Development in the United States: Results from the NLSY." *Children and Youth Services Review* 17, no. 1 (1995): 127–55.

Kowal, Emma. "The Promise of Indigenous Epigenetics." *Discover Society*, October 4, 2016. https://discoversociety.org/2016/10/04/the-promise-of-indigenous-epigenetics/.

Kramer, M. S., and K. S. Joseph. "Enigma of Fetal/Infant-Origins Hypothesis." *Lancet* 348, no. 9037 (1996): 1254–55.

Kretchmer, Norman. "Ecology of the Newborn Infant." In *The Infant at Risk: Early Detection and Preventive Intervention*, edited by Daniel Bergsma. New York: Intercontinental Medical Book Corp, 1974.

Krieger, Nancy. "Embodiment: A Conceptual Glossary for Epidemiology." *Journal of Epidemiology and Community Health* 59, no. 5 (2005): 350–55.

Kristoff, Nicholas D. "At Risk from the Womb." *New York Times*, October 2, 2010.

Kuechenhoff, Bernhard. "The Psychiatrist Auguste Forel and His Attitude to Eugenics." *History of Psychiatry* 19, no. 2 (2008): 215–23.

Kuh, Diana, and Yoav Ben-Shlomo. "Should We Intervene to Improve Fetal and Infant Growth?" In *A Life Course Approach to Chronic Disease Epidemiology*. Oxford: Oxford University Press, 2004.

Kukla, Quill Rebecca. *Mass Hysteria: Medicine, Culture, and Mothers' Bodies*. Lanham, MD: Rowman & Littlefield, 2005.

Kuzawa, C. *Workshop on the Maternal Environment: Chris Kuzawa on the Plasticity of Fetal Life*. Chicago: Human Capital and Economic Opportunity Global Working Group, 2016.

Kuzawa, C. "Fetal Origins of Developmental Plasticity: Are Fetal Cues Reliable Predictors of Future Nutritional Environments?" *American Journal of Human Biology* 17, no. 1 (2005): 5–21.

Kuzawa, C. "Why Evolution Needs Development, and Medicine Needs Evolution." *International Journal of Epidemiology* 41, no. 1 (2012): 223–29.

Kuzawa, C., P. D. Gluckman, and M. A. Hanson. "Developmental Perspectives on the Origins of Obesity." In *Adipose Tissue and Adipokines in Health and Disease*, edited by E. Fantuzzi and T. Mazzone. Totowa, NJ: Humana Press, 2016.

Kuzawa, C., and Elizabeth Sweet. "Epigenetics and the Embodiment of Race: Developmental Origins of US Racial Disparities in Cardiovascular Health." *American Journal of Human Biology* 21, no. 1 (2009): 2–15.

Ladd-Taylor, Molly. *Mother-Work: Women, Child Welfare, and the State, 1890–1930*. Urbana: University of Illinois Press, 1994.

Ladd-Taylor, Molly. "Eugenics, Sterilisation and Modern Marriage in the USA: The Strange Career of Paul Popenoe." *Gender & History* 13, no. 2 (2001): 298–327.

Lamarck, Jean-Baptiste. *Zoological Philosophy: An Exposition with Regard to the Natural History of Animals*. Chicago: University of Chicago Press, 1984.

Lamoreaux, Janelle. "What If the Environment Is a Person? Lineages of Epigenetic Science in a Toxic China." *Cultural Anthropology* 31, no. 2 (2016): 188–214.

Lampl, Michelle. "Obituary for Professor David Barker." *Annals of Human Biology* 41, no. 2 (2014): 187–90.

Landecker, Hannah. "Food as Exposure: Nutritional Epigenetics and the New Metabolism." *BioSocieties* 6, no. 2 (2011): 167–94.

Landecker, Hannah, and Aaron Panofsky. "From Social Structure to Gene Regulation, and Back: A Critical Introduction to Environmental Epigenetics for Sociology." *Annual Review of Sociology* 39, no. 1 (2013): 333–57.

Landrigan, Philip J., Joel Forman, Maida Galvez, Brooke Newman, Stephanie M. Engel, and Claude Chemtob. "Impact of September 11 World Trade Center Disaster on Children and Pregnant Women." *Mount Sinai Journal of Medicine: A Journal of Translational and Personalized Medicine* 75, no. 2 (2008): 129–34.

Langley, K., J. Heron, G. D. Smith, and A. Thapar. "Maternal and Paternal Smoking during Pregnancy and Risk of ADHD Symptoms in Offspring: Testing for Intrauterine Effects." *American Journal of Epidemiology* 176, no. 3 (2012): 261–68.

Langley, K., F. Rice, M. B. van den Bree, and A. Thapar. "Maternal Smoking during Pregnancy as an Environmental Risk Factor for Attention Deficit Hyperactivity Disorder Behaviour: A Review." *Minerva Pediatrica* 57, no. 6 (2005): 359–71.

Langley-Evans, Simon, A. Swali, and Sarah McMullen. "Lessons from Animal Models: Mechanisms of Nutritional Programming." In *Early Life Nutrition, Adult Health and Development: Lessons from Changing Diets, Famines and Experimental Studies*, edited by L. H. Lumey and Alexander Vaiserman, 253–80. New York: Nova Biomedical, 2013.

Lappé, Martine. "The Paradox of Care in Behavioral Epigenetics: Constructing Early-Life Adversity in the Lab." *BioSocieties* 13, no. 4 (2018): 698–714.

Larregue, Julien, and Oliver Rollins. "Biosocial Criminology and the Mismeasure of Race." *Ethnic and Racial Studies* 42, no. 12 (2019): 1990–2007.

Lassi, Zohra S., Philippa F. Middleton, Caroline Crowther, and Zulfiqar A. Bhutta. "Interventions to Improve Neonatal Health and Later Survival: An Overview of Systematic Reviews." *EBioMedicine* 2, no. 8 (2015): 985–1000.

Laubichler, Manfred D. "Hertwig, Wilhelm August Oscar." In *Encyclopedia of Life Sciences*, 617–18, 2005.

Laubichler, Manfred D., and Eric H. Davidson. "Boveri's Long Experiment: Sea Urchin Merogones and the Establishment of the Role of Nuclear Chromosomes in Development." *Developmental Biology* 314, no. 1 (2008): 1–11.

Lauderdale, Diane S. "Birth Outcomes for Arabic-Named Women in California before and after September 11." *Demography* 43, no. 1 (2006): 185–201.

Lawlor, D. A. "The Society for Social Medicine John Pemberton Lecture 2011. Developmental Overnutrition: An Old Hypothesis with New Importance?" *International Journal of Epidemiology* 42, no. 1 (2013): 7–29.

Lawlor, D. A., G. D. Batty, S. M. Morton, I. J. Deary, S. Macintyre, G. Ronalds, and D. A. Leon. "Early Life Predictors of Childhood Intelligence: Evidence from the Aberdeen Children of the 1950s Study." *Journal of Epidemiology and Community Health* 59, no. 8 (2005): 656–63.

Lawlor, D. A., C. Relton, N. Sattar, and S. M. Nelson. "Maternal Adiposity: A Determinant of Perinatal and Offspring Outcomes?" *Nature Reviews Endocrinology* 8, no. 11 (2012): 679–88.

Lawlor, D. A., N. J. Timpson, R. M. Harbord, S. Leary, A. Ness, M. I. McCarthy, T. M. Frayling, A. T. Hattersley, and G. D. Smith. "Exploring the Developmental Overnutrition Hypothesis Using Parental-Offspring Associations and FTO as an Instrumental Variable." *PLOS Med* 5, no. 3 (2008): e33.

Lee, David. "Forel, Auguste-Henri." In *Encyclopedia of the History of Psychological Theories*, 446–47, 2012.

Lee, T. M., and I. Zucker. "Vole Infant Development Is Influenced Perinatally by Maternal Photoperiodic History." *American Journal of Physiology—Regulatory, Integrative and Comparative Physiology* 255, no. 5 (1988): R831–38.

Leeuwenhoeck, Anthony. "An Abstract of a Letter from Mr. Anthony Leeuwenhoeck of Delft about Generation by an Animalcule of the Male Seed. Animals in the Seed of a Frog. Some Other Observables in the Parts of a Frog. Digestion, and the Motion of the Blood in a Feavor." *Philosophical Transactions* 13, no. 143–154 (1683): 347–55.

Lei, Man-Kit, Steven R. H. Beach, Ronald L. Simons, and Robert A. Philibert. "Neighborhood

Crime and Depressive Symptoms among African American Women: Genetic Moderation and Epigenetic Mediation of Effects." *Social Science & Medicine* 146 (2015): 120–28.

Lev, Katy Rank. "Why You Should Worry about Grandma's Eating Habits." *Mother Nature News*, 2014. http://www.mnn.com/health/fitness-well-being/stories/why-you-should -worry-about-grandmas-eating-habits.

Lewis, Sophie. *Full Surrogacy Now: Feminism against Family*. London: Verso, 2019.

Lewontin, Richard C. *The Triple Helix: Gene, Organism, and Environment*. Cambridge, MA: Harvard University Press, 2000.

Library of Congress, Copyright Office, Catalogues of Copyright Entries. New Series. Vol. 9. Washington, DC: Government Printing Office, 1913.

Lieberman, Adam. "Video Display Terminals, 1989." In *Facts versus Fears: A Review of the Greatest Unfounded Health Scares of Recent Times*, 40–42. New York: American Council on Science and Health, 2004.

Little, R. E. "Mother's and Father's Birthweight as Predictors of Infant Birthweight." *Paediatric and Perinatal Epidemiology* 1, no. 1 (1987): 19–31.

Lloyd, Stéphanie, and Eugene Raikhel. "'It Was There All Along': Situated Uncertainty and the Politics of Publication in Environmental Epigenetics." *BioSocieties* 13, no. 4 (2018): 737–60.

Loeb, Jacques. *The Organism as a Whole, from a Physicochemical Viewpoint*. New York: Putnam's Sons, 1916.

Lombrozo, Tania. "Causal-Explanatory Pluralism: How Intentions, Functions, and Mechanisms Influence Causal Ascriptions." *Cognitive Psychology* 61, no. 4 (2010): 303–32.

Lombrozo, Tania. "Using Science to Blame Mothers." *NPR Cosmos & Culture: Commentary on Science and Society*, August 25, 2014. https://www.npr.org/sections/13.7/2014/08/25/ 343121679/using-science-to-blame-mothers-check-your-values.

Longino, Helen E. *Studying Human Behavior: How Scientists Investigate Aggression and Sexuality*. Chicago: University of Chicago Press, 2013.

Los Angeles Herald. "Prevention Is Their Object, Mrs. Mary E. Teats Talks of Purity Work, National Lecturer Holds First Series of Meetings." November 19, 1901, 11.

Loudon, Irvine. *Death in Childbirth: An International Study of Maternal Care and Maternal Mortality, 1800–1950*. Oxford: Oxford University Press, 1992.

Lowe, C. R. "Effect of Mothers' Smoking Habits on Birth Weight of Their Children." *BMJ* 2, no. 5153 (1959): 673–76.

Lu, Michael, and Jessica Chow. "An Interview with David Barker." ca. 2014. http://www.lcrn .net/an-interview-with-david-barker/.

Lucas, A. "Programming by Early Nutrition in Man." *Ciba Foundation Symposium* 156 (1991): 38–50.

Lucas, A. "Role of Nutritional Programming in Determining Adult Morbidity." *Archives of Disease in Childhood* 71, no. 4 (1994): 288–90.

Lumey, L. H., A. D. Stein, H. S. Kahn, K. M. van der Pal-de Bruin, G. J. Blauw, P. A. Zybert, and E. S. Susser. "Cohort Profile: The Dutch Hunger Winter Families Study." *International Journal of Epidemiology* 36, no. 6 (2007): 1196–204.

Lumey, L. H., Aryeh D. Stein, and Ezra Susser. "Prenatal Famine and Adult Health." *Annual Review of Public Health* 32, no. 1 (2011): 237–62.

Lumey, L. H., and Alexander Vaiserman, eds. *Early Life Nutrition, Adult Health and Development: Lessons from Changing Dietary Patterns, Famines and Experimental Studies*. New York: Nova Biomedical, 2013.

Lumley, J., and L. Donohue. "Aiming to Increase Birth Weight: A Randomised Trial of Pre-pregnancy Information, Advice and Counselling in Inner-Urban Melbourne." *BMC Public Health* 6, no. 1 (2006): 299.

Maccauley, T. B. "The Supposed Inferiority of First and Second Born Members of Families: Statistical Fallacies." *Journal of Heredity* 2, no. 3 (1911): 165–75.

Macdorman, Marian F. "Race and Ethnic Disparities in Fetal Mortality, Preterm Birth, and Infant Mortality in the United States: An Overview." *Seminars in Perinatology* 35, no. 4 (2011): 200–208.

Macfadden, Bernarr. *Manhood and Marriage.* New York: Physical Culture Publishing, 1916.

Machamer, Peter, Lindley Darden, and Carl F. Craver. "Thinking about Mechanisms." *Philosophy of Science* 67, no. 1 (2000): 1–25.

MacMahon, Brian, Marc Alpert, and Eva J. Salber. "Infant Weight and Parental Smoking Habits." *American Journal of Epidemiology* 82, no. 3 (1965): 247–61.

Malabou, Catherine. "One Life Only: Biological Resistance, Political Resistance." *Critical Inquiry*, 2015. http://criticalinquiry.uchicago.edu/one_life_only/.

Mamluk, L., H. B. Edwards, J. Savovic, V. Leach, T. Jones, T. H. M. Moore, S. Ijaz, et al. "Low Alcohol Consumption and Pregnancy and Childhood Outcomes: Time to Change Guidelines Indicating Apparently 'Safe' Levels of Alcohol during Pregnancy? A Systematic Review and Meta-Analyses." *BMJ Open* 7, no. 7 (2017): e015410.

Markens, Susan, C. H. Browner, and Nancy Press. "Feeding the Fetus: On Interrogating the Notion of Maternal-Fetal Conflict." *Feminist Studies* 23, no. 2 (1997): 351–72.

Marshall, C. F. *Syphilology and Venereal Disease.* 3rd ed. New York: W. Wood, 1914.

Martin, Emily. "The Egg and the Sperm: How Science Has Constructed a Romance Based on Stereotypical Male-Female Roles." *Signs* 16, no. 3 (1991): 485–501.

Martin, Nina, and Nina Montagne. "The Last Person You'd Expect to Die in Childbirth." *ProPublica*, 2017. https://www.propublica.org/article/die-in-childbirth-maternal-death-rate-health-care-system-1.

"Maternal Impressions—Belief in Their Existence Is Due to Unscientific Method of Thought—No Evidence Whatever That Justifies Faith in Them—How the Superstition Originated." *Journal of Heredity* 6 (1915): 512–18.

Maupas, Emile. "Le rajeunissement karyogamique chez les Cilies." *Archives de zoologie expérimentale et générale, 2 ser.* 7, nos. 1, 2, 3 (1889).

McBride, W. G. "Thalidomide and Congenital Abnormalities." *Lancet* 278 (1961): 1358.

McGanity, W. J., E. B. Bridgforth, M. P. Martin, J. A. Newbill, and W. J. Darby. "The Vanderbilt Cooperative Study of Maternal and Infant Nutrition, VIII: Some Nutritional Implications." *Journal of the American Dietetic Association* 31, no. 6 (1955): 582.

McGill, Ann L. "Context Effects in Judgments of Causation." *Journal of Personality and Social Psychology* 57, no. 2 (1989): 189–200.

McLaren, Angus. "Medium and Message." *Journal of Modern History* 46, no. 1 (1974): 86–97.

McLaren, Anne, and Donald Michie. "An Effect of the Uterine Environment upon Skeletal Morphology in the Mouse." *Nature*, no. 4616 (1958): 1147–48.

McLaren, Anne, and Donald Michie. "Factors Affecting Vertebral Variation in Mice, 4: Experimental Proof of the Uterine Basis of a Maternal Effect." *Journal of Embryology and Experimental Morphology* 6, no. 4 (1958): 645–59.

McPhail, Deborah, Andrea Bombak, Pamela Ward, and Jill Allison. "Wombs at Risk, Wombs as Risk: Fat Women's Experiences of Reproductive Care." *Fat Studies* 5, no. 2 (2016): 98–115.

Melendy, Mary Ries. *Perfect Womanhood for Maidens, Wives, Mothers.* Chicago: Monarch, 1903.

Melendy, Mary Ries. *The Science of Eugenics and Sex Life: Sex-Life, Love, Marriage, Maternity.* Harrisburg, PA: Minter, 1914.

Melendy, Mary Ries. *Vivilore: The Pathway to Mental and Physical Perfection.* Chicago: W. R. Vansant, 1904.

Meloni, Maurizio. *Political Biology: Science and Social Values in Human Heredity from Eugenics to Epigenetics.* New York: Palgrave Macmillan, 2016. doi:40025877678.

Meloni, Maurizio. "A Postgenomic Body: Histories, Genealogy, Politics." *Body & Society* 24, no. 3 (2018): 3–38.

Meyer, Mary B., and George W. Comstock. "Maternal Cigarette Smoking and Perinatal Mortality." *American Journal of Epidemiology* 96, no. 1 (1972): 1–10.

Michelson, Nicholas. "Studies in the Physical Development of Negroes, II: Weight." *American Journal of Physical Anthropology* 1, no. 3 (1943): 289–300.

Mill, J., and B. T. Heijmans. "From Promises to Practical Strategies in Epigenetic Epidemiology." *Nature Reviews Genetics* 14, no. 8 (2013): 585–94.

Milner, Kate M., Trevor Duke, and Ingrid Bucens. "Reducing Newborn Mortality in the Asia-Pacific Region: Quality Hospital Services and Community-Based Care." *Journal of Paediatrics and Child Health* 49, no. 7 (2013): 511–18.

Montagu, Ashley. *The Biosocial Nature of Man.* New York: Grove Press, 1956.

Montagu, Ashley. *Life before Birth.* New York: New American Library, 1964.

Montagu, Ashley. *Prenatal Influences.* Springfield, IL: C. C. Thomas, 1962.

Morgan, Hugh D., Heidi G. E. Sutherland, David I. K. Martin, and Emma Whitelaw. "Epigenetic Inheritance at the Agouti Locus in the Mouse." *Nature Genetics* 23, no. 3 (1999): 314.

Morgan, Thomas Hunt. *The Physical Basis of Heredity.* Philadelphia: J. B. Lippincott, 1919.

Morrow, Prince A. *Social Diseases and Marriage, Social Prophylaxis.* New York: Lea Brothers, 1904.

Morton, J. S. "Riddell, Newton N." In *Illustrated History of Nebraska: A History of Nebraska from the Earliest Explorations of the Trans-Mississippi Region,* edited by Albert Watkins, 559. Lincoln: Western Publishing and Engraving Company, 1913.

Morton, Newton E. "The Inheritance of Human Birth Weight." *Annals of Human Genetics* 20, no. 2 (1955): 125–34.

Mosher, Martha B. *Child Culture in the Home: A Book for Mothers.* New York: F. H. Revell Company, 1898.

Mosley, Michael. "Feeling Stressed? Then Blame Your Mother." *London Times,* September 28, 2015.

Mott, F. W. *Nature and Nurture in Mental Development.* London: J. Murray, 1914.

Moynihan, Daniel P. *The Negro Family: The Case for National Action.* Washington, DC: United States Department of Labor, 1965.

M.R & Anor, An tArd Chlaraitheoir & Ors, [2013] IEHC 91 Abbott J. (judgment, High Court of Ireland).

Müller, R. "A Task That Remains before Us: Reconsidering Inheritance as a Biosocial Phenomenon." *Seminars in Cell and Developmental Biology* 97 (2020): 189–94.

Müller, Ruth, and Georgia Samaras. "Epigenetics and Aging Research: Between Adult Malleability and Early Life Programming." *BioSocieties* 13, no. 4 (2018): 715–36.

Murphy, Michelle. "Distributed Reproduction, Chemical Violence, and Latency." *Scholar & Feminist Online* 11, no. 3 (2013). http://sfonline.barnard.edu/life-un-ltd-feminism-bioscience-race/distributed-reproduction-chemical-violence-and-latency/.

Murphy, Michelle. *The Economization of Life.* Durham: Duke University Press, 2017.

Murphy, T. M., and J. Mill. "Epigenetics in Health and Disease: Heralding the EWAS Era." *Lancet* 383, no. 9933 (2014): 1952–54.

Naeye, R. L., M. M. Diener, and W. S. Dellinger. "Urban Poverty: Effects on Prenatal Nutrition." *Science* 166, no. 3908 (1969): 1026.

Newnham, J. P., and M. G. Ross. *Early Life Origins of Human Health and Disease.* Basel: Karger, 2009.

"News and Notes." *Texas School Journal* 24, no. 9 (1907): 30.

Newton, A. E. *Pre-natal Culture: Being Suggestions to Parents Relative to Systemic Methods of Moulding the Tendencies of Offspring before Birth.* Washington, DC: Moral Education Society, 1879.

New York Times. "A Woman's Rights: Parts 1–7." December 28, 2018.

Niewöhner, Jörg. "Epigenetics: Embedded Bodies and the Molecularisation of Biography and Milieu." *BioSocieties* 6, no. 3 (2011): 279–98.

Oken, E., A. A. Baccarelli, D. R. Gold, K. P. Kleinman, A. A. Litonjua, D. De Meo, J. W. Rich-Edwards, et al. "Cohort Profile: Project Viva." *International Journal of Epidemiology* 44, no. 1 (2015): 37–48.

Olby, Robert C. "Constitutional and Hereditary Disorders." In *Companion Encyclopedia of the History of Medicine,* edited by W. F. Bynum and Roy Porter. New York: Routledge, 1993.

Oregon Humanities. "Epigenetics and Equity: Zip Code May Be More Important than Genetic Code When It Comes to Determining a Person's Health." Edited by Dan Sadowsky. 2014. https://www.oregonhumanities.org/rll/beyond-the-margins/epigenetics-and-equity/.

Osiro, S., J. Gielecki, P. Matusz, M. M. Shoja, R. S. Tubbs, and M. Loukas. "August Forel (1848–1931): A Look at His Life and Work." *Child's Nervous System* 28, no. 1 (2012): 1–5.

Osofsky, Howard. "Poverty, Pregnancy Outcome, and Child Development." In *The Infant at Risk: Early Detection and Preventive Intervention,* edited by Daniel Bergsma. New York: Intercontinental Medical Book Corp, 1974.

Oster, Emily. *Expecting Better: Why the Conventional Wisdom Is Wrong—and What You Really Need to Know.* New York: Penguin Press, 2013.

Ounsted, Margaret, and Christopher Ounsted. "Maternal Regulation of Intra-uterine Growth." *Nature* 212, no. 5066 (1966): 995–97.

Ounsted, Margaret, and Christopher Ounsted. *On Fetal Growth Rate: Its Variations and Their Consequences.* Clinics in Developmental Medicine. Vol. 46. London: Heinemann Medical; J. B. Lippincott for Spastics International Medical Publications, 1973.

Overy, C., L. A. Reynolds, and E. M. Tansey, eds. *History of the Avon Longitudinal Study of Parents and Children (ALSPAC), c.1980–2000.* Vol. 44. Wellcome Witnesses to Twentieth Century Medicine. London: Queen Mary, University of London, 2012.

"An Overstretched Hypothesis?" *Lancet* 357, no. 9254 (2001): 405.

Owen, Richard. *On Parthenogenesis.* London: J. Van Voorst, 1849.

Oyama, Susan. *The Ontogeny of Information: Developmental Systems and Evolution.* Cambridge: Cambridge University Press, 1985.

Pacchierotti, F., and M. Spano. "Environmental Impact on DNA Methylation in the Germline: State of the Art and Gaps of Knowledge." *BioMed Research International* 2015 (2015): 123484.

Palloni, A., and J. D. Morenoff. "Interpreting the Paradoxical in the Hispanic Paradox: Demographic and Epidemiologic Approaches." *Annals of the New York Academy of Science* 954 (2001): 140–74.

Paltrow, Lynn. "Criminal Prosecutions against Pregnant Women." In *Reproductive Freedom Project*. New York: American Civil Liberties Union Foundation, 1992.

Paltrow, L. M., and J. Flavin. "Arrests of and Forced Interventions on Pregnant Women in the United States, 1973–2005: Implications for Women's Legal Status and Public Health." *Journal of Health Politics, Policy, and Law* 38, no. 2 (2013): 299–343.

Pandora, Katherine. "Knowledge Held in Common: Tales of Luther Burbank and Science in the American Vernacular." *Isis* 92, no. 3 (2001): 484.

Paneth, N., and M. Susser. "Early Origin of Coronary Heart Disease (the 'Barker Hypothesis')." *BMJ* 310, no. 6977 (1995): 411–12.

Panofsky, Aaron. *Misbehaving Science: Controversy and the Development of Behavior Genetics.* Chicago: University of Chicago Press, 2014.

Paré, Ambroise, and Janis L. Pallister. *On Monsters and Marvels.* Chicago: University of Chicago Press, 1982.

Park, Katharine. *Secrets of Women: Gender, Generation, and the Origins of Human Dissection.* New York: Zone Books, 2006.

Parker, George. "Mothers at Large: Responsibilizing the Pregnant Self for the 'Obesity Epidemic.'" *Fat Studies* 3, no. 2 (2014): 101–18.

Parker, George. "Shamed into Health? Fat Pregnant Women's Views on Obesity Management Strategies in Maternity Care." *Women's Studies Journal* 31, no. 1 (2017): 22–33.

Parker, George, and Cat Pausé. "'I'm Just a Woman Having a Baby': Negotiating and Resisting the Problematization of Pregnancy Fatness." *Frontiers in Sociology* 3 (2018).

Paul, Annie Murphy. *Origins: How the Nine Months before Birth Shape the Rest of Our Lives.* New York: Free Press, 2010.

Paul, Diane B. *Controlling Human Heredity, 1865 to the Present.* Atlantic Highlands, NJ: Humanities Press, 1995.

Payne, Jenny Gunnarsson. "Grammars of Kinship: Biological Motherhood and Assisted Reproduction in the Age of Epigenetics." *Signs: Journal of Women in Culture and Society* 41, no. 3 (2016): 483–506.

Peitzman, Steven J. "Forgotten Reformers: The American Academy of Medicine." *Bulletin of the History of Medicine* 58, no. 4 (1984): 516–29.

Pendleton, Hester. *The Parents Guide for the Transmission of Desired Qualities to Offspring, and Childbirth Made Easy.* Stereotype ed. New York: Fowlers and Wells, 1848.

Perkins Gilman, Charlotte. "How Home Conditions React Upon the Family." *American Journal of Sociology* 14, no. 1 (1909): 592–605.

Peterson, Erik L. *The Life Organic: The Theoretical Biology Club and the Roots of Epigenetics.* Pittsburgh: University of Pittsburgh Press, 2016.

Pollitt, Katha. "Fetal Rights." In *"Bad" Mothers: The Politics of Blame in Twentieth-Century America*, edited by Molly Ladd-Taylor and Lauri Umansky. New York: New York University Press, 1998.

Popenoe, Paul, and Roswell Hill Johnson. *Applied Eugenics.* New York: Macmillan, 1918.

"Prevention of Neural Tube Defects: Results of the Medical Research Council Vitamin Study. MRC Vitamin Study Research Group." *Lancet* 338, no. 8760 (1991): 131–37.

"Prof. Oscar Hertwig." *Nature* 111, no. 2776 (1923): 56.

Purdy, Laura M. "Are Pregnant Women Fetal Containers?" *Bioethics* 4, no. 4 (1990): 273–91.

Purnell, Beverly A. "The Nutritional Sins of the Mother . . ." *Science* 345, no. 6198 (2014): 782.

Rader, Karen A. *Making Mice: Standardizing Animals for American Biomedical Research, 1900–1955*. Princeton: Princeton University Press, 2004.

Raine, Adrian. *The Anatomy of Violence: The Biological Roots of Crime*. New York: Pantheon Books, 2013.

Rasmussen, Kathleen Maher. "The 'Fetal Origins' Hypothesis: Challenges and Opportunities for Maternal and Child Nutrition." *Annual Review of Nutrition* 21, no. 1 (2001): 73–95.

Ravelli, G. P., Z. A. Stein, and M. W. Susser. "Obesity in Young Men after Famine Exposure in Utero and Early Infancy." *New England Journal of Medicine* 295, no. 7 (1976): 349–53.

Ray, L. Bryan. "Inheriting Mom's Exercise Regime." *Science* 352, no. 6284 (2016): 425–26.

Read, Mary Lillian. *The Mothercraft Manual*. Boston: Little, Brown, 1916.

Reardon, Sara. "Poverty Linked to Epigenetic Changes and Mental Illness." *Nature*. May 24, 2016. https://www.nature.com/news/poverty-linked-to-epigenetic-changes-and-mental-illness-1.19972.

Reiches, M. W. "A Life History Approach to Prenatal Supplementation: Building a Bridge from Biological Anthropology to Public Health and Nutrition." *American Journal of Human Biology* 31, no. 6 (2019): e23318.

"Report of the National Woman's Christian Temperance Union Thirty-Fifth Annual Convention." Denver, Colorado. 1909.

Reynolds, Gretchen. "Babies Born to Run: Does Exercise during Pregnancy Lead to Exercise-Loving Offspring?" *New York Times*, April 12, 2016.

Rhoads, J. E. "Mrs. Teats at San Jose." *Pacific Ensign* 15, no. 13 (1905): 3.

Richardson, Sarah S. "Maternal Bodies in the Postgenomic Order: Gender and the Explanatory Landscape of Epigenetics." In *Postgenomics: Perspectives on Biology after the Genome*, edited by Sarah S. Richardson and Hallam Stevens, 210–31. Durham: Duke University Press, 2015.

Richardson, Sarah S. "Plasticity and Programming: Feminism and the Epigenetic Imaginary." *Signs* 43, no. 1 (2017): 29–52.

Richardson, Sarah S., and Rene Almeling. "The CDC Risks Its Credibility with New Pregnancy Guidelines." *Boston Globe*, February 8, 2016. https://www.bostonglobe.com/opinion/2016/02/08/the-cdc-risks-its-credibility-with-new-pregnancy-guidelines/2SCHzNCqcWNDRguol7kzwK/story.html.

Richardson, S. S., C. R. Daniels, M. W. Gillman, J. Golden, R. Kukla, C. Kuzawa, and J. Rich-Edwards. "Society: Don't Blame the Mothers." *Nature* 512, no. 7513 (2014): 131–32.

Richardson, Sarah S., and Hallam Stevens. *Postgenomics: Perspectives on Biology after the Genome*. Durham: Duke University Press, 2015.

Richmond, R. C., C. L. Relton, and G. Davey Smith. "What Evidence Is Required to Suggest that DNA Methylation Mediates the Association between Prenatal Famine Exposure and Adulthood Disease?" *Science Advances* 4, no. 1 (January 31, 2018): eaao4364.

Richmond, R. C., N. J. Timpson, J. F. Felix, T. Palmer, R. Gaillard, G. McMahon, G. Davey Smith, V. W. Jaddoe, and D. A. Lawlor. "Using Genetic Variation to Explore the Causal Effect of Maternal Pregnancy Adiposity on Future Offspring Adiposity: A Mendelian Randomisation Study." *PLOS Med* 14, no. 1 (2017): e1002221.

Riddell, Newton N. *Child Culture According to the Laws of Physiological Psychology and Mental Suggestion*. Chicago: Child of Light Publishing, 1902.

Riddell, Newton N. *A Child of Light, or, Heredity and Prenatal Culture: Considered in the Light of the New Psychology*. Chicago: Child of Light Publishing, 1900.

Riddell, Newton N. "Experiments of Elmer Gates." In *The Riddell Lectures on Applied Psychology and Vital Christianity*. Chicago: Riddell Publishers, 1913.

Riddell, Newton N. "Suggestions Relating to Children." *Kindergarten-Primary Magazine* 26, no. 8 (1914): 226.

Rijlaarsdam, J., C. A. Cecil, E. Walton, M. S. Mesirow, C. L. Relton, T. R. Gaunt, W. McArdle, and E. D. Barker. "Prenatal Unhealthy Diet, Insulin-Like Growth Factor 2 Gene (IGF2) Methylation, and Attention Deficit Hyperactivity Disorder Symptoms in Youth with Early-Onset Conduct Problems." *Journal of Child Psychology and Psychiatry* 58, no. 1 (2017): 19–27.

Rijlaarsdam, J., I. Pappa, E. Walton, M. J. Bakermans-Kranenburg, V. R. Mileva-Seitz, R. C. Rippe, S. J. Roza, et al. "An Epigenome-Wide Association Meta-analysis of Prenatal Maternal Stress in Neonates: A Model Approach for Replication." *Epigenetics* 11, no. 2 (2016): 140–49.

Roberts, Dorothy E. "The Ethics of Biosocial Science." The Tanner Lectures on Human Values, 2016. tannerlectures.utah.edu.

Roberts, Dorothy E. *Fatal Invention: How Science, Politics, and Big Business Re-create Race in the Twenty-First Century*. New York: New Press, 2011.

Roberts, Dorothy E. *Killing the Black Body: Race, Reproduction, and the Meaning of Liberty*. 1st ed. New York: Pantheon Books, 1997.

Robison, O. W. "The Role of Maternal Effects in Animal Breeding, V: Maternal Effects in Swine." *Journal of Animal Science* 35, no. 6 (1972): 1303–15.

Rock Island Argus. "Psychology of Heredity Theme, N. N. Riddell's Lecture on Prenatal Culture Best Yet Given; Cites Many Experiments." February 3, 1912, 6.

Rodwell, Grant. "Dr Caleb Williams Saleeby: The Complete Eugenicist." *History of Education* 26, no. 1 (1997): 23–40.

Roizen, Michael F., and Mehmet Oz. *YOU: Having a Baby: The Owner's Manual to a Happy and Healthy Pregnancy*. New York: Free Press, 2009.

Roseboom, T. J., R. C. Painter, A. F. van Abeelen, M. V. Veenendaal, and S. R. de Rooij. "Hungry in the Womb: What Are the Consequences? Lessons from the Dutch Famine." *Maturitas* 70, no. 2 (2011): 141–45.

Rosen, Christine. *Preaching Eugenics: Religious Leaders and the American Eugenics Movement*. New York: Oxford University Press, 2004.

Rosenberg, Charles E. "The Bitter Fruit: Heredity, Disease and Social Thought in Nineteenth Century America." *Perspectives in American History* 8 (1974): 189–235.

Rosner, Elizabeth. *Survivor Café: The Legacy of Trauma and the Labyrinth of Memory*. Berkeley: Counterpoint, 2017.

Rundle, Andrew, Aryeh D. Stein, Henry S. Kahn, Karin van der Pal-de Bruin, Patricia A. Zybert, and L. H. Lumey. "Anthropometric Measures in Middle Age after Exposure to Famine during Gestation: Evidence from the Dutch Famine." *American Journal of Clinical Nutrition* 85, no. 3 (2007): 869–76.

Rush, D., Z. Stein, and M. Susser. "A Randomized Controlled Trial of Prenatal Nutritional Supplementation in New York City." *Pediatrics* 65, no. 4 (1980): 683–97.

Russell Sage Foundation. "Call for Proposals: Integrating Biology and Social Science Knowledge (BioSS)." https://www.russellsage.org/research/funding/bioss.

Russett, Cynthia Eagle. *Sexual Science: The Victorian Construction of Womanhood*. Cambridge, MA: Harvard University Press, 1989.

Saldaña-Tejeda, Abril. "Mitochondrial Mothers of a Fat Nation: Race, Gender and Epigenetics in Obesity Research on Mexican Mestizos." *BioSocieties* 13 (2017).

Saleeby, C. W. *Parenthood and Race Culture: An Outline of Eugenics*. New York: Moffat, Yard, 1909.

Sapp, Jan. *Beyond the Gene: Cytoplasmic Inheritance and the Struggle for Authority in Genetics*. New York: Oxford University Press, 1987.

Schneider, William H. "Puericulture, and the Style of French Eugenics." *History and Philosophy of the Life Sciences* 8, no. 2 (1986): 265–77.

Schuller, Kyla C. "Taxonomies of Feeling: The Epistemology of Sentimentalism in Late-Nineteenth-Century Racial and Sexual Science." *American Quarterly* 64, no. 2 (2012): 277–99.

Scott, Cameron. "Link between Mom's Diet at Conception and Child's Lifelong Health." *SingularityHub*. May 9, 2014. https://singularityhub.com/2014/05/09/link-between-moms-diet-at-conception-and-childs-lifelong-health-study-reveals/#sm.00001k28ging79elttqo9hoqcbxsh.

Scott, R. B., W. W. Cardozo, A. deG. Smith, and M. R. DeLilly. "Growth and Development of Negro Infants, III: Growth during the First Year of Life as Observed in Private Pediatric Practice." *Journal of Pediatrics* 37, no. 6 (1950): 885–93.

Scott, R. B., M. E. Jenkins, and R. P. Crawford. "Growth and Development of Negro Infants, I: Analysis of Birth Weights of 11,818 Newly Born Infants." *Pediatrics* 6, no. 3 (1950): 425–31.

"Section Discussions: Superstition in Teratology." *JAMA* 48, no. 4 (1907): 363–64.

Semon, Richard. *The Mneme*. London: George Allen & Unwin, 1921.

Shannon, T. W., and W. J. Truitt. *Nature's Secrets Revealed; Scientific Knowledge of the Laws of Sex Life and Heredity, or Eugenics*. Marietta, OH: S. A. Mullikin, 1914.

Shapiro, Sam, Harold Jacobziner, Paul M. Densen, and Louis Weiner. "Further Observations on Prematurity and Perinatal Mortality in a General Population and in the Population of a Prepaid Group Practice Medical Care Plan." *American Journal of Public Health and the Nation's Health* 50, no. 9 (1960): 1304–17.

Sharp, Gemma C., Deborah A. Lawlor, and Sarah S. Richardson. "It's the Mother! How Assumptions about the Causal Primacy of Maternal Effects Influence Research on the Developmental Origins of Health and Disease." *Social Science & Medicine* 213 (2018): 20–27.

Sharp, G. C., L. Schellhas, S. S. Richardson, and D. A. Lawlor. "Time to Cut the Cord: Recognizing and Addressing the Imbalance of DOHaD Research towards the Study of Maternal Pregnancy Exposures." *Journal of Developmental Origins of Health and Disease* 10, no. 5 (2019): 509–12.

Shelly, Edwin Taylor. "Superstition in Teratology with Special Reference to the Theory of Impressionism." *JAMA* 48, no. 4 (1907): 308–11.

Shildrick, Margrit. "Maternal Imagination: Reconceiving First Impressions." *Rethinking History* 4, no. 3: 243–60.

Shipley, Arthur Everett, Edward Bagnall Poulton, and Selmar Schönland. "Editors' Preface to the First Edition." In *Essays upon Heredity and Kindred Biological Problems*, vii. Oxford: Clarendon Press, 1891.

Shostak, Sara, and Margot Moinester. "The Missing Piece of the Puzzle? Measuring the Environment in the Postgenomic Moment." In *Postgenomics: Perspectives on Biology after the Genome*, edited by Sarah S. Richardson and Hallam Stevens, 192–209. Durham, NC: Duke University Press, 2015.

Shulevitz, Judith. "The Science of Suffering: Kids Are Inheriting Their Parents' Trauma. Can Science Stop It?" *New Republic*, November 16, 2014.

Silverstein, Jason. "How Racism Is Bad for Our Bodies." *Atlantic*, March 12, 2013. https://www.theatlantic.com/health/archive/2013/03/how-racism-is-bad-for-our-bodies/273911/.

Sizer, Nelson. "Child Culture: Specimens of Promise." *Phrenological Journal and Science of Health* 102, no. 8 (1896): 60.

Slater, William K., and J. Edwards. "John Hammond, 1889–1964." *Biographical Memoirs of Fellows of the Royal Society* 11 (1965): 101–13.

Slemons, Josiah Morris. *The Prospective Mother: A Handbook for Women during Pregnancy.* New York: D. Appleton, 1912.

Slocum, Annette. *For Wife and Mother: A Young Mother's Tokology.* Chicago: Medical Publishing, 1910.

Smith, Clement A. "Effects of Maternal Undernutrition upon the Newborn Infant in Holland (1944–1945)." *Journal of Pediatrics* 30, no. 3 (1947): 229–43.

Smith, Hannah Whitall. *Child Culture; Or, the Science of Motherhood.* New York: Fleming H. Revell, 1894.

Smith, Justin E. H. "Imagination and the Problem of Heredity in Mechanist Embryology." In *The Problem of Animal Generation in Early Modern Philosophy*, edited by Justin E. H. Smith, 80–99. Cambridge: Cambridge University Press, 2006.

Spicer, Kirsten. "'A Nation of Imbeciles': The Human Betterment Foundation's Propaganda for Eugenic Practices in California." *Voces Novae: Chapman University Historical Review* 7, no. 1 (2015): 109–30.

Squires, Sally. "The Tiny Infants: National Academy of Sciences Maps Strategy to Prevent Underweight Births." *Washington Post*, February 27, 1985.

Stacpoole, Florence. *Advice to Women on the Care of the Health, before, during and after Confinement.* Rev. from the 5th London ed. New York: Funk & Wagnalls, 1917.

Stein, Aryeh D., Patricia A. Zybert, Karin van der Pal-de Bruin, and L. H. Lumey. "Exposure to Famine during Gestation, Size at Birth, and Blood Pressure at age 59 y: Evidence from the Dutch Famine." *European Journal of Epidemiology* 21, no. 10 (2006): 759–65.

Stein, Zena, Mervyn Susser, Gerhart Saenger, and Francis Marolla. *Famine and Human Development: The Dutch Hunger Winter of 1944–1945.* New York: Oxford University Press, 1975.

Stern, Madeleine B. *Heads and Headlines: The Phrenological Fowlers.* Norman: University of Oklahoma Press, 1971.

Stopes, Marie Carmichael. "Statement of Dr. Marie Stopes." In *Problems of Population and Parenthood*, edited by James Marchant, 242–55. New York: E. P. Dutton, 1920.

Stotz, Karola. "The Ingredients for a Postgenomic Synthesis of Nature and Nurture." *Philosophical Psychology* 21, no. 3 (2008): 359–81.

Strahan, S. A. K. *Marriage and Disease: A Study of Heredity and the More Important Family Degenerations.* London: Kegan Paul, Trench, Trübner, 1892.

Sturtevant, A. H. "Inheritance of Direction of Coiling in *Limnaea*." *Science* 58, no. 1501 (1923): 269–70.

Suter, Melissa, Jun Ma, Alan S. Harris, Lauren Patterson, Kathleen A. Brown, Cynthia Shope, Lori Showalter, Adi Abramovici, and Kjersti M. Aagaard-Tillery. "Maternal Tobacco Use Modestly Alters Correlated Epigenome-Wide Placental DNA Methylation and Gene Expression." *Epigenetics* 6, no. 11 (2011): 1284–94.

Szyf, M., I. C. Weaver, F. A. Champagne, J. Diorio, and M. J. Meaney. "Maternal Programming of Steroid Receptor Expression and Phenotype through DNA Methylation in the Rat." *Frontiers in Neuroendocrinology* 26, no. 3–4 (2005): 139–62.

Tanaka, Yoshimaro. "Maternal Inheritance in *Bombyx mori*." *Genetics* 9, no. 5 (1924): 479–86.

Tanner, J. M. "Standards for Birth Weight or Intra-uterine Growth." *Pediatrics* 46, no. 1 (1970): 1–6.

Tanner, J. M. *A History of the Study of Human Growth.* New York: Cambridge University Press, 1981.

Teats, Mary E. *The Way of God in Marriage: A Series of Essays upon Gospel and Scientific Purity.* Spotswood, NJ: Physical Culture, 1906.

Terao, Hajime. "Maternal Inheritance in the Soy Bean." *American Naturalist* 52, no. 613 (1918): 51–56.

Thomson, A. M. "Diet in Pregnancy, 3: Diet in Relation to the Course and Outcome of Pregnancy." *British Journal of Nutrition* 13, no. 4 (1959): 509–25.

Tobi, Elmar W., P. Eline Slagboom, Jenny van Dongen, Dennis Kremer, Aryeh D. Stein, Hein Putter, Bastiaan T. Heijmans, and L. H. Lumey. "Prenatal Famine and Genetic Variation Are Independently and Additively Associated with DNA Methylation at Regulatory Loci within IGF2/H19." *PLOS One* 7, no. 5 (2012): e37933.

Tobi, Elmar W., Roderick C. Slieker, René Luijk, Koen F. Dekkers, Aryeh D. Stein, Kate M. Xu, P. Eline Slagboom, et al. "DNA Methylation as a Mediator of the Association between Prenatal Adversity and Risk Factors for Metabolic Disease in Adulthood." *Science Advances* 4, no. 1 (2018): eaao4364.

Tobi, Elmar W., R. C. Slieker, A. D. Stein, H. E. Suchiman, P. E. Slagboom, E. W. van Zwet, B. T. Heijmans, and L. H. Lumey. "Early Gestation as the Critical Time-Window for Changes in the Prenatal Environment to Affect the Adult Human Blood Methylome." *International Journal of Epidemiology* 44, no. 4 (2015): 1211–23.

Tomlinson, Stephen. "Phrenology, Education and the Politics of Human Nature: The Thought and Influence of George Combe." *History of Education* 26, no. 1 (1997): 1–22.

Towbin, Abraham. "Mental Retardation Due to Germinal Matrix Infarction." *Science* 164, no. 3876 (1969): 156–61.

Toyama, Kametaro. "Maternal Inheritance and Mendelism." *Journal of Genetics* 2, no. 4 (1913): 351–405.

Tran, Hoang T., Lex W. Doyle, Katherine J. Lee, and Stephen M. Graham. "A Systematic Review of the Burden of Neonatal Mortality and Morbidity in the ASEAN Region." *WHO South-East Asia Journal of Public Health* 1, no. 3 (2012): 239.

"Truth, Reconciliation, and Transformation: Continuing on the Path to Equity." *Pediatrics* 146, no. 3 (2020): e2020019794.

Uda, Hajime. "On 'Maternal Inheritance.'" *Genetics* 8, no. 4 (1923): 322–35.

Udry, J. Richard, Naomi M. Morris, Karl E. Bauman, and Charles L. Chase. "Social Class, Social Mobility, and Prematurity: A Test of the Childhood Environment Hypothesis for Negro Women." *Journal of Health and Social Behavior* 11, no. 3 (1970): 190–95.

Uncle, Joseph. "Joyous and Sensitive." *Phrenological Journal and Science of Health* 107, no. 5 (1899): 158.

Unger, J. "Weight at Birth and Its Effect on Survival of the Newborn by Geographic Divisions and Urban Rural Areas: United States, Early 1950." *Vital and Health Statistics.* Series 21, Data from the National Vital Statistics System, no. 4 (1965): 155–218.

"Use of Folic Acid for Prevention of Spina Bifida and Other Neural Tube Defects—1983–1991." *MMWR Morbidity and Mortality Weekly Report* 40, no. 30 (1991): 513–16.

Valdez, Natali. "The Redistribution of Reproductive Responsibility: On the Epigenetics of 'Environment' in Prenatal Interventions." *Medical Anthropology Quarterly* 32, no. 3 (2018).

VanderWeele, Tyler J. *Explanation in Causal Inference: Methods for Mediation and Interaction.* New York: Oxford University Press, 2015.

van Wyhe, John. "The Diffusion of Phrenology through Public Lecturing." In *Science in the*

Marketplace: Nineteenth-Century Sites and Experiences, edited by Aileen Fyfe and Bernard Lightman, 60. Chicago: University of Chicago Press, 2007.

Venge, Ole. "Studies of the Maternal Influence on the Birth Weight in Rabbits." *Acta Zoologica* 31, no. 1 (1950): 1–148.

Villarosa, Linda. "Why America's Black Mothers and Babies Are in a Life-or-Death Crisis." *New York Times,* April 11, 2018.

Viterna, J., and J. S. G. Bautista. "Pregnancy and the 40-Year Prison Sentence: How 'Abortion Is Murder' Became Institutionalized in the Salvadoran Judicial System." *Health and Human Rights* 19, no. 1 (2017): 81–93.

Waggoner, Miranda R. *The Zero Trimester: Pre-pregnancy Care and the Politics of Reproductive Risk.* Oakland: University of California Press, 2017.

Waggoner, Miranda R., and T. Uller. "Epigenetic Determinism in Science and Society." *New Genetics and Society* 34, no. 2 (2015): 177–95.

Wald, Lillian D., Jane Addams, Leo Arnstein, Ben B. Lindsey, Henry B. Favill, Charles R. Henderson, Florence Kelley, and Samuel Mccune Lindsay. "The Federal Children's Bureau: A Symposium." *Annals of the American Academy of Political and Social Science* 33 (1909): 23–48.

Walton, Arthur, and John Hammond. "The Maternal Effects on Growth and Conformation in Shire Horse-Shetland Pony Crosses." *Proceedings of the Royal Society of London, Series B—Biological Sciences* 125, no. 840 (1938): 311–35.

Warin, Megan, and Anne Hammarström. "Material Feminism and Epigenetics: A 'Critical Window' for Engagement?" *Australian Feminist Studies* 33, no. 97 (2018): 299–315.

Warin, Megan, V. Moore, T. Zivkovic, and M. Davies. "Telescoping the Origins of Obesity to Women's Bodies: How Gender Inequalities Are Being Squeezed Out of Barker's Hypothesis." *Annals of Human Biology* 38, no. 4 (2011): 453–60.

Warin, Megan, Tanya Zivkovic, Vivienne Moore, and Michael Davies. "Mothers as Smoking Guns: Fetal Overnutrition and the Reproduction of Obesity." *Feminism & Psychology* 22, no. 3 (2012): 360–75.

Warkany, Josef. "Experimental Studies on Nutrition in Pregnancy." *Obstetrical & Gynecological Survey* 3, no. 5 (1948): 693–703.

Warkany, Josef. "Manifestations of Prenatal Nutritional Deficiency." In *Vitamins and Hormones III,* 73–103. New York: Academic Press, 1945.

Warkany, Josef, and Harold Kalter. "Congenital Malformations." *New England Journal of Medicine* 265, no. 21 (1961): 1046–52.

Washington Post. "Art of Brain Building: Prof. Elmer Gates and His Remarkable Experiments." April 5, 1896.

Waterland, R. A., D. C. Dolinoy, J. R. Lin, C. A. Smith, X. Shi, and K. G. Tahiliani. "Maternal Methyl Supplements Increase Offspring DNA Methylation at Axin Fused." *Genesis* 44, no. 9 (2006): 401–6.

Waterland, R. A., and R. L. Jirtle. "Transposable Elements: Targets for Early Nutritional Effects on Epigenetic Gene Regulation." *Molecular Cell Biology* 23, no. 15 (2003): 5293–300.

Weaver, Ian C. G., Nadia Cervoni, Frances A. Champagne, Ana C. D'Alessio, Shakti Sharma, Jonathan R. Seckl, Sergiy Dymov, Moshe Szyf, and Michael J. Meaney. "Epigenetic Programming by Maternal Behavior." *Nature Neuroscience* 7, no. 8 (2004): 847–54.

Weaver, Ian C. G., Michael J. Meaney, and Moshe Szyf. "Maternal Care Effects on the Hippocampal Transcriptome and Anxiety-Mediated Behaviors in the Offspring That Are Reversible in Adulthood." *Proceedings of the National Academy of Sciences* 103, no. 9 (2006): 3480–85.

Weaver, Lawrence T. "In the Balance: Weighing Babies and the Birth of the Infant Welfare Clinic." *Bulletin of the History of Medicine* 84 (2010): 30–57.

Weismann, August. "Amphimixis or the Essential Meaning of Conjugation and Sexual Reproduction (1891)." In *Essays upon Heredity and Kindred Biological Problems*, edited by Edward Bagnall Poulton, S. Schönland, and A. E. Shipley. Oxford: Clarendon Press, 1891.

Weismann, August. "Beiträge zur Naturgeschichte der Daphnoiden. Abhandlung VI und VII." *Zeitschrift für Wissenschaftliche Zoologie*, no. Erstes und Zweites Heft (1880): 55.

Weismann, August. "The Continuity of the Germ-Plasm as the Foundation of a Theory of Heredity (1885)." In *Essays upon Heredity and Kindred Biological Problems*, edited by Edward Bagnall Poulton, S. Schönland, and A. E. Shipley. Oxford: Clarendon, 1891.

Weismann, August. *The Germ-Plasm: A Theory of Heredity*. New York: Scribner's, 1893.

Weismann, August. "On Heredity (1883)." In *Essays upon Heredity and Kindred Biological Problems*, edited by Edward Bagnall Poulton, S. Schönland, and A. E. Shipley. Oxford: Clarendon, 1891.

Weismann, August. "Remarks on Certain Problems of the Day (1890)." In *Essays upon Heredity and Kindred Biological Problems*, edited by Edward Bagnall Poulton, S. Schönland, and A. E. Shipley. Oxford: Clarendon, 1891.

Weismann, August. "The Significance of Sexual Reproduction in the Theory of Natural Selection (1886)." In *Essays upon Heredity and Kindred Biological Problems*, edited by Edward Bagnall Poulton, S. Schönland, and A. E. Shipley. Oxford: Clarendon, 1891.

Weismann, August. "The Supposed Transmission of Mutilations (1888)." In *Essays upon Heredity and Kindred Biological Problems*, edited by Edward Bagnall Poulton, S. Schönland, and A. E. Shipley. Oxford: Clarendon, 1891.

Weismann, August, Edward Bagnall Poulton, S. Schönland, and A. E. Shipley. *Essays upon Heredity and Kindred Biological Problems*. 2nd ed. 2 vols. Oxford: Clarendon, 1891.

"Well-Born Children: The Slogan of the Twentieth Century!! Correspondence School of Gospel and Scientific Eugenics." Chicago, 1911. Author's collection.

Wells, Jonathan C. K. "Maternal Capital and the Metabolic Ghetto: An Evolutionary Perspective on the Transgenerational Basis of Health Inequalities." *American Journal of Human Biology* 22, no. 1 (2010): 1–17.

Wells, Jonathan C. K. *The Metabolic Ghetto: An Evolutionary Perspective on Nutrition, Power Relations and Chronic Disease*. Cambridge: Cambridge University Press, 2016.

Wertz, Richard W., and Dorothy C. Wertz. *Lying-In: A History of Childbirth in America*. New York: Free Press, 1977.

West, Mary Mills. *Prenatal Care*. US Children's Bureau Care of Children Series. 2nd ed. Vol. 4. Washington, DC: Government Printing Office, 1913.

Whitehead, James. *On the Transmission, from Parent to Offspring, of Some Forms of Disease, and of Morbid Taints and Tendencies*. 2nd ed. London: J. Churchill, 1857.

Wilcox, Allen J. "On the Importance—and the Unimportance—of Birthweight." *International Journal of Epidemiology* 30, no. 6 (2001): 1233–41.

Willham, R. L. "The Role of Maternal Effects in Animal Breeding." *Journal of Animal Science* 35 (1972): 1288–93.

Wilson, Edmund B. *The Cell in Development and Inheritance*. London: MacMillan, 1896.

Wilson, Edmund B. *The Cell in Development and Heredity*. 3rd ed. New York: Macmillan, [1925] 1928.

Winett, Liana B., Alyssa B. Wulf, and Lawrence Wallack. "Framing Strategies to Avoid Mother-

Blame in Communicating the Origins of Chronic Disease." *American Journal of Public Health* 106, no. 8 (2016): 1369–73.

"Winfield Scott Hall 1861–1942." *Quarterly Bulletin of Northwestern University Medical School* 16, no. 4 (1942): 315.

Wingerter, Rex B. "Fetal Protection Becomes Assault on Motherhood." *In These Times*, June 10, 1987, 3, 8.

Winick, Myron. "Food and the Fetus." *Natural History* 90, no. 1 (1981): 76–81.

Winick, Myron, ed. *Nutrition and Fetal Development*. Vol. 1, *Current Concepts in Nutrition*. New York: Wiley, 1974.

Winter, Alison. *Mesmerized: Powers of Mind in Victorian Britain*. Chicago: University of Chicago Press, 1998.

Winther, Rasmus G. "August Weismann on Germ-Plasm Variation." *Journal of the History of Biology* 34, no. 3 (2001): 517–55.

Wolf, Jason B., and Michael J. Wade. "What Are Maternal Effects (and What Are They Not)?" *Philosophical Transactions of the Royal Society B: Biological Sciences* 364, no. 1520 (2009): 1107–15.

Wolf, Joan B. *Is Breast Best? Taking on the Breastfeeding Experts and the New High Stakes of Motherhood*. New York: New York University Press, 2011.

Woodward, James. "Causation in Biology: Stability, Specificity, and the Choice of Levels of Explanation." *Biology & Philosophy* 25, no. 3 (2010): 287–318.

Woodward, James. *Making Things Happen: A Theory of Causal Explanation*. New York: Oxford University Press, 2003.

World Health Organization. "Obesity and Overweight." *Fact Sheets*, 2018.

Wright, John, Neil Small, Pauline Raynor, Derek Tuffnell, Raj Bhopal, Noel Cameron, Lesley Fairley, et al.; on behalf of the Born in Bradford Scientific Collaborators Group. "Cohort Profile: The Born in Bradford Multi-ethnic Family Cohort Study." *International Journal of Epidemiology* 42, no. 4 (2012): 978–91.

Yehuda, R., and L. M. Bierer. "Transgenerational Transmission of Cortisol and PTSD Risk." *Progress in Brain Research* 167 (2008): 121–35.

Yehuda, R., N. P. Daskalakis, L. M. Bierer, H. N. Bader, T. Klengel, F. Holsboer, and E. B. Binder. "Holocaust Exposure Induced Intergenerational Effects on FKBP5 Methylation." *Biological Psychiatry* 80, no. 5 (2016): 372–80.

Yerby, A. S. "The Disadvantaged and Health Care." *American Journal of Public Health and the Nation's Health* 56, no. 1 (1966): 5–9.

Yerushalmy, J. "Nomenclature, Duration of Gestation, Birth Weight, Intrauterine Growth: Dissenting Views." *Pediatrics* 39, no. 6 (1967): 940–41.

Yoshizawa, Rebecca Scott. "Fetal-Maternal Intra-action: Politics of New Placental Biologies." *Body and Society* 22, no. 4 (2016): 79–105.

Young, Dwight L. "Orson Squire Fowler: To Form a More Perfect Human." *Wilson Quarterly* 14, no. 2 (1990): 120–27.

Yu, Z., S. Han, J. Zhu, X. Sun, C. Ji, and X. Guo. "Pre-pregnancy Body Mass Index in Relation to Infant Birth Weight and Offspring Overweight/Obesity: A Systematic Review and Meta-analysis." *PLOS One* 8, no. 4 (2013): e61627.

Zamenhof, S., and E. Van Marthens. "Study of Factors Influencing Prenatal Brain Development." *Molecular and Cellular Biochemistry* 4, no. 3 (1974): 157–68.

Zenderland, Leila. "Biblical Biology: American Protestant Social Reformers and the Early Eugenics Movement." *Science in Context* 11, no. 3–4 (1998).

Index

Page numbers in italics refer to figures and tables.

25, 127; Weismann on, 111, 116–17, 218; womb and, 110, 119, 197, 209–10, 214, 258n48

Maternal Impressions (Bayer), 62, 74

maternal impressions theories: Dixon on, 84–85; fetal programming and, 20; germ plasm and, 84–87, 91, 105, 245n6; heredity and, 27, 34–39, 48, 54, 236n27; obstetrics and, 81–82, 84; prenatal culture and, 57–67, 73–82, 241n32; superstition and, 21, 48, 64, 79, 82, 84–87, 105; traits and, 21, 34, 39, 58, 74, 87; Weismann and, 20–21, 37, 39, 48, 54, 58, 60, 79–80

maternal imprint: acquired traits and, 34–39; alcohol and, 14–15, 17–18, 21; BeginBeforeBirth.org and, 11–13; biosocial body and, 135, 160; birth weight and, 6, 14, 22–23; causality and, 4–11, 23–24; constitution of identity and, 234n31; criminality and, 11–15; Developmental Origins of Health and Disease (DOHaD) and, 3, 5, 10, 12, 16–17, 21, 24; diet and, 8, 10, 12–13, 18, 20; DNA and, 2, 4, 7–10; emotion and, 13, 34, 208; epidemiology and, 3, 12, 22; epigenetics and, 4, 10–13, 17, 23–25, 199, 214; eugenics and, 20–21; fetal origins science and, 3–5, 12; fetal programming and, 3, 5, 7–8, 17, 168–69, 180, 184–85; fetus and, 2, 5–9, 13–21; gender and, 2, 19, 25; germ plasm and, 103; gestation and, 1, 7, 18–20, 196; Golschmidt on, 116, 118; heredity and, 103 (*see also* heredity); historical perspective on, 13–16, 19–25, 109, 116, 118, 220, 237n45; Holocaust and, 1–2; inheritance and, 7, 17, 21–22, 24; intensive mothering and, 16–19, 235n58; intrauterine environment and, 1–7, 10–11, 13, 17, 19–23; metabolism and, 5, 17; methylation and, 23; nutrition and, 3, 5–6, 9, 20; obesity and, 3–5, 17–18, 215; phenotype and, 2, 7, 22; pregnancy and, 2–25, 235n58; prenatal culture and, 57, 76; psychiatry and, 1, 3, 11; smoking and, 13–15; stress and, 1, 3, 5–6, 9, 11–13, 17; Sturtevant on, 115; transgenerational,

225; trauma and, 1–2, 7, 10; Uda on, 114–15; uterus and, 12, 19; Weismann on, 20–25; women as fetal containers and, 13–16

Maupas, Emile, 41–42, 46

McBean, Gordon, 180

McGill, Ann, 203

McGill University, 222

McLaren, Anne, 53, 125–26, 131

Meaney, Michael, 222

Mears, Martha, 244n159

measles, 137

Meharry Medical College, 141

meiosis, 33, 40

Melendy, Mary R., 56, 64, 65, 73–74, 77

Meloni, Maurizio, 223

Mendelian theory, 28, 151; Bateson and, 110–11, 116, 131, 239n110; Biffen and, 110–11, 114; Dobzhansky and, 110, 119, 121, 131; inheritance and, 22, 110–15, 117; Loeb on, 117–18; maternal effects and, 108–18; traits and, 22, 49, 61, 109–10, 117–18, 239n110; Uda on, 114–15

Mendel's Principles of Heredity (Bateson), 110

menstruation, 34

mental retardation, 73, 153

Metabolic Ghetto, The (Wells), 5

metabolism: fetal programming and, 161–62, 164, 169, 177, 179–80, 182, 186, 190–94; heredity and, 35, 222; maternal effects and, 117, 214–15; maternal imprint and, 5, 17

methylation: candidate genes and, *163–67*, 173–74; CpGs and, *163–66*, 172, *175–76*, *184*, 192–94; diet and, 177, 191, 200, 256n73; fetal programming and, 161–95; FKBP5 gene and, *167*, 184–89; "gene knockout" experiment and, 199–200; heredity and, 221–22; IGF2 and, *163*, 191–92, 256n73; Infinium Human Methylation 450k Beadchip microarray and, *174*, *176*, *179*; maternal effects and, 198–202, 209, 217, 258n48; maternal imprint and, 23; trauma and, *187*, *193*; whole genome population and, *163–66*, 173–74, *176*, *179*, 192, 194

mice, 22, 31, 59, 120, 124–27, 138, 172, 177, 197–200, 249n58